Historical-Analytical Studies on Nature, Mind and Action

Volume 2

Historical-Analytical Studies on Nature, Mind and Action provides a forum for integrative, multidisciplinary, analytic studies in the areas of philosophy of nature, philosophical anthropology, and the philosophy of mind and action in their social setting. Tackling these subject areas from both a historical and contemporary systematic perspective, this approach allows for various "paradigm-straddlers" to come together under a common umbrella. Digging down to the conceptual-historical roots of contemporary problems, one will inevitably find common strands which have since branched out into isolated disciplines. This series seeks to fill the void for studies that reach beyond their own strictly defined boundaries not only synchronically (reaching out to contemporary disciplines), but also diachronically, by investigating the unquestioned contemporary presumptions of their own discipline by taking a look at the historical development of those presumptions and the key concepts they involve. This series, providing a common forum for this sort of research in a wide range of disciplines, is designed to work against the well-known phenomenon of disciplinary isolation by seeking answers to our fundamental questions of the human condition: What is there? -- What can we know about it? -- What should we do about it? – indicated by the three key-words in the series title: Nature, Mind and Action. This series will publish monographs, edited volumes, revised doctoral theses and translations.

More information about this series at http://www.springer.com/series/11934

Miguel García-Valdecasas • José Ignacio Murillo
Nathaniel F. Barrett
Editors

Biology and Subjectivity

Philosophical Contributions to Non-reductive Neuroscience

Springer

Editors
Miguel García-Valdecasas
Mind-Brain Group, Institute for Culture and Society (ICS)
University of Navarra
Pamplona, Spain

José Ignacio Murillo
Mind-Brain Group, Institute for Culture and Society (ICS)
University of Navarra
Pamplona, Spain

Nathaniel F. Barrett
Mind-Brain Group, Institute for Culture and Society (ICS)
University of Navarra
Pamplona, Spain

ISSN 2509-4793 ISSN 2509-4807 (electronic)
Historical-Analytical Studies on Nature, Mind and Action
ISBN 978-3-319-30501-1 ISBN 978-3-319-30502-8 (eBook)
DOI 10.1007/978-3-319-30502-8

Library of Congress Control Number: 2016949065

Printed on acid-free paper

This Springer imprint is published by Springer Nature
The registered company is Springer International Publishing AG Switzerland

This book was completed with the help and support of the Institute for Culture and Society of the University of Navarra (Spain)

ics
Universidad
de Navarra

Contents

Chapter 1
Biology and Subjectivity: Philosophical Contributions to a Non-reductive Neuroscience

José Ignacio Murillo, Miguel García-Valdecasas, and Nathaniel F. Barrett

In the middle of the twentieth century, Wittgenstein warned that "the method of reducing the explanation of natural phenomena to the smallest possible number of primitive natural laws…leads…into complete darkness" (1958, p. 18). At the time, few philosophers and even fewer scientists were prepared to heed his warning. A half-century later, however, the reductive method of science—the method famously defined by Descartes, brilliantly exemplified by Newtonian physics, and long upheld as the gold standard of scientific explanation—seems to have finally lost its luster. While reduction is still widely defended, in the last decades alternative views have gained credibility, to the extent that a "non-reductive science" is no longer dismissed as an oxymoron.

This change is partly due to failures of reductive science. Most prominent of these is the failure of physics to produce a "grand unifying theory" that explains all natural phenomena using a few mathematical formulae. In response, a number of prominent physicists have called for a new approach with different explanatory standards and goals (Wolfram 2002; Laughlin 2005; Smolin 2006). Similarly, despite the "neuro-hype" of recent decades, a leading neuroscientist has recently claimed that "we currently have plenty of knowledge about the 'how' of the brain but still lack an answer to the 'what' of the brain. We thus remain blind to its main and overarching purpose" (Northoff 2013, p. xi). More positively, however, the success of innovative approaches in biology and various fields devoted to the study of mind indicates the promise of non-reductive science: witness the notable examples of Paul Weiss (1973), Robert Rosen (1991), Francisco Varela (2000), and Stuart Kauffman (2000) in biology; and J.A. Scott Kelso (1995), Walter Freeman (1999),

J.I. Murillo • M. García-Valdecasas (✉) • N.F. Barrett
Mind-Brain Group, Institute for Culture and Society (ICS), University of Navarra, 31009 Pamplona, Spain
e-mail: jimurillo@unav.es; garciaval@unav.es; nbarrett@unav.es

M. García-Valdecasas et al. (eds.), *Biology and Subjectivity*, Historical-Analytical Studies on Nature, Mind and Action 2,
DOI 10.1007/978-3-319-30502-8_1

Maxwell Bennett (Bennett and Hacker 2003) and Georg Northoff (2014) in neuroscience.

These examples indicate that for the scientific understanding of living systems, but especially highly intelligent living systems such as ourselves, multiple levels of explanation seem to be indispensable, and not just because their complexity limits our understanding of their constituent parts. Rather, an adequate understanding of the behavior of the parts of a living system seems to depend on the dynamic organization of the system as a whole, and this organization, in turn, seems to depend on the purposeful activity of the system within its environment.

The title of this volume, "Biology and subjectivity," is intended to highlight the close connection between a more adequate, non-reductive understanding of mind with similarly non-reductive understanding of life. For the views presented here, the latter may not be sufficient for the former, but it is at least necessary. But more importantly, going in the other direction, the contributions of this volume seek to demonstrate how careful reflection on subjectivity is necessary for an adequate understanding of life in general.

Our use of the term "subjectivity" is broadly inclusive. It covers feeling, affectivity, value, intentionality, and more—in short, all of the traits of human intelligence and experience that have seemed resistant to scientific explanation and, for this reason, have so often been reduced or "explained away" by standard scientific models. A common assumption in the last century has been that the human mind presents an intractable explanatory "Hard Problem," while the rest of nature is relatively non-problematic for traditional scientific approaches. In contrast, many of this volume's essays are decidedly more optimistic about the possibility of making explanatory progress concerning human subjectivity, provided we adopt a different approach to living systems: one that incorporates subjectivity into biology.

Of course, a non-reductive approach that involves multiple levels of dynamic organization as well as purposiveness and other features of subjectivity is a theory that needs to be borne out by scientific investigation. Its development will likely involve the adoption of some of the same methods and the consideration of the same data that were developed and harvested under the auspices of more restrictive explanatory frameworks. So we should be careful not to portray non-reductive science as if it makes a clean break with "traditional" reductive science. As with all theoretical gains, the transition from reductive to non-reductive science is contingent upon showing that old data can be convincingly interpreted by new theories.

On the other hand, the transition to a non-reductive framework may require much more than the careful reinterpretation of data. Depending on the object of investigation and the questions at hand, the relationship between theory and data can become so entangled that revisions of a much more fundamental nature are required. The apparent necessity of such large-scale revisions would seem to militate against one of the most noteworthy philosophical statements of non-reductive neuroscience to date, Bennett and Hacker's *Philosophical Foundations of Neuroscience* (2003). This work introduces a rather neat and tidy division of labor between philosophy and science, corresponding to a sharp distinction between conceptual and empirical questions. That is, while philosophy addresses conceptual questions, science deals with empirical questions. Although we agree with many of Bennett and Hacker's analyses of the conceptual flaws of neuroscientific theories (e.g. the mereological fallacy), we do not share such a rigid division of labor. This volume has been assembled with the intent to explore the possibility that a genuinely non-reductive science of the mind calls for a more complex and involved relationship between philosophy and science. A relevant example will help to clarify this point.

In a recent review (2009), Mazviita Chirimuuta and Ian Gold call into question the long-standing "classical" view of the receptive field (RF) of neurons in the primary visual cortex (V1) in a manner that indirectly raises the issue of explanatory reduction. As established by the Nobel prize-winning research of Hubel and Wiesel (1959), the classical view is that neurons of the visual system have fixed RFs, such that each neuron of the V1 responds to a specific pattern of stimuli. Experimental evidence for fixed RFs—though never undisputed (see below)—has lent empirical support to one of the most influential theoretical and methodological principles of neuroscience, the "neuron doctrine" (Barlow 1972; Guillery 2005; Bullock et al. 2005). This is the idea that individual neurons constitute the basic functional "building blocks" of perceptual and cognitive processes and that investigations of neural systems should first determine the properties of individual neurons and build up from there. When applied in this way, the neuron doctrine is a clear example of the traditional Cartesian approach to scientific understanding: in the case of vision, for instance, the expectation is that adequate knowledge of the components and architecture of the visual system will allow us to assemble a working model of the system as whole. A crucial assumption for this approach is that basic functional properties of the components—e.g., the RFs of individual neurons—are relatively fixed, that is, unaffected by the activity of neurons at the same or higher levels of the visual pathways. However, recent research in V1 physiology has not supported this assumption (ibid., pp. 207-12); instead, it points to a substantially revised, dynamic notion of the RF or perhaps even to the rejection of the RF concept altogether (pp. 214–15). The response properties of V1 neurons seem to vary depending on the nature of stimuli—lighting conditions, patterns of motion—and other conditions, including perhaps the arousal and interest of the perceiving subject. Accordingly, one possibility that arises from this research is that an adequate understanding of our visual system requires reference to the dynamic influences of higher levels of circuitry, which themselves may be dependent on the purposive activity of the animal within a natural environment.

Now, what can we infer about the role of philosophy from this potentially dramatic turn in the neuroscience of vision? Chirimuuta and Gold are rather circumspect in their assessment of its implications. They refuse to draw any conclusions about the classical model of vision, let alone the neuron doctrine. Rather, they argue that questions about RFs are to be settled empirically by further investigation and, moreover, that the best way to settle these questions is to continue to explore revised versions of classical "single-unit" explanations of visual processes along with "circuit-level" explanations (pp. 216–17). Philosophers who are eager to move past reductive frameworks would do well to keep their example of prudence in mind. On the other hand, Chirimuuta and Gold's statement that "the encouragement of philosophy is neither necessary nor particularly helpful" (p. 217) seems to be contradicted by their observation that the failure of scientists to pay attention to the "complex nonlinear properties of RFs" was only partly due to the limited availability of data. Chirimuuta and Gold cite a 1953 paper that notes the "flexibility and fluidity" of RF activity patterns (Kuffler 1953; cited in Chirimuuta and Gold, p. 211). The implication is that neuroscientists such as Hubel and Wiesel could have adopted a very different theoretical approach, one that embraced a dynamic, situation-dependent view of neural response properties. If they had done so, perhaps empirical research of the last half-century would have proceeded very differently (see Noë 2009, pp. 149–69). In light of this oversight, together with the fact that many philosophers have long argued for a more interactive model of perception (e.g. Dewey 1972 [1896]), cannot we imagine a more constructive role for philosophy in its engagement with science?

This question cannot be deflected by saying that philosophical complexities only get in the way of scientific progress—namely, the practical business of devising testable models and carrying out experiments. The history of science shows that it is naïve to think that any theory that yields readily testable predictions is worth pursuing. Scientific theories are not so easily falsified, especially once they are established within the scientific community. As long as there is new data to be mined, a "degenerate research programme" (Lakatos 1970) can persist for decades despite diminishing explanatory returns. Also, we should be wary of the idea that scientific fields progress through stages of increasing complexity, such that simplification is a necessary first step toward understanding. Surely Chirimuuta and Gold are right that "the most promising route for any new science has always been to seek out any underlying simplicity in what appears to be a formidably complex and unpredictable object of investigation" (p. 212). But this begs a key question: *What kind of simplicity should we be looking for?*

As exemplified by the neuron doctrine, the guiding assumption of modern science has been that the simplicity we seek will take the form of basic building blocks from which we can assemble models of more complex phenomena. In recent decades, however, this assumption has been increasingly called into question in a wide variety of fields by scientists and philosophers alike. Such broad questioning of reductionism cannot be viewed as just another paradigm shift. We are not just deciding between theories or even between paradigms; we are talking about the very meaning of scientific explanation. Perhaps, then, the bias that leads scientists to believe that the only way to progress is to break down objects of investigation into Cartesian building blocks stems from the deep acceptance of some rather abstract ideas about causality, identity, and the ultimate nature of science. If so, this bias may not be removed without richer engagement between philosophy and science.

Fortunately, we are not without exemplars for this difficult task. Scholars in a wide range of disciplines have begun to engage critically with neuroscience with the aim of promoting a more integrated understanding of human behavior (Choudhury and Slaby 2012). Among philosophers, Maxine Sheets-Johnstone (2011), Alicia Juarrero (1999), Shaun Gallagher (2006), Alva Noë (2009), Evan Thompson (2007), Nancy Murphy (2006), Dan Zahavi (2005), and P.M.S. Hacker (Bennett and

Hacker2003) have demonstrated the value of engagement with neuroscience, illuminating the significance of recent scientific advances while, at the same time, deflating scientific claims to have "explained" mind and consciousness by brain processes alone. These critical and constructive exchanges belie the simplistic view that philosophy must accept the findings of science without reflecting on the assumptions that underpin scientific analysis, including the assumption that common experience, the wellspring of philosophy, is inferior to "hard data" as a source of knowledge, or that only scientific data provides a proper basis for philosophical theories.

Still, there is no easy way to define the proper stance of philosophical work whose aim is to contribute to a non-reductive science of mind. Engagement with ongoing research has to be combined with a certain degree of imaginative "distancing" from dominant explanatory frameworks such as computational or information-processing theories of mind. In this respect, philosophers who have embraced a reductive stance (e.g. Bickle 2003) have a more straightforward, if not easier, task, as they typically do not have to imagine how current research would look from the perspective of an entirely different framework. The risk of such a close partnership with science is that philosophers can become overly invested in theories that fall in and out of favor over the course of a decade or so. Perhaps it is better for philosophers to take the long view and remain somewhat aloof from the shifting currents of scientific research. On the other hand, if philosophical contributions are to be taken seriously by the scientific community, philosophers must be willing to make their theories vulnerable to correction by scientific experimentation.

In our understanding, one of the philosophers who better exemplified the intrinsic relation between philosophy and science is Aristotle. Perhaps no other figure in western thought enjoys the stature of progenitor of both speculative philosophy and empirical science, even if his contribution to science, while impressive, was unsystematic. While not every chapter in this volume deals specifically with Aristotle's views or draws from the Aristotelian tradition, all of them can be said to be Aristotelian in a broad sense: they adopt a non-reductive and empirically committed approach to life and mind. In this sense, an Aristotelian approach can be described as any empirically guided search for explanatory principles that are appropriate to the kind of phenomena to be explained. Such an approach calls for the sophisticated combination of theory and observation usually attributed to Aristotle. The same thinker who gave us the grand systematic vision of the *Metaphysics* and *Categories* is also considered to be the first biologist and even the first physiological psychologist. In the words of Daniel Robinson, a historian of psychology:

> No predecessor could possibly or plausibly lay claim to the title of an early physiological psychologist, and this is precisely the title we may assign to Aristotle. He was the first authority to delineate a domain specifically embracing the subject matter of psychology, and within that domain, to confine his explanations to principles of a biological sort. That the entire body of Aristotelian philosophy does not fit into a materialist mold is clear; the philosopher himself goes to some lengths to make it clear. But on the narrower issues of learning, memory, sleep and dreams, routine perception, animal behavior, emotion, and motivation, Aristotle's approach is naturalistic, psychological, and empirical (1976, p. 83; cited in Keeley 2009, pp. 228–29).

In fact, Aristotle's legacy for modern psychology is rather complicated, especially with respect to the issue of reductionism. This is indicated by recent debates over the question of whether or not Aristotle's psychology counts as an ancient precursor to functionalism (Nussbaum and Rorty 1992; Langton 2000), which has been portrayed as a kind of non-reductive approach to mind. Although we cannot enter into the details of these debates here, it is worth noting the irony that the same philosopher who is celebrated as the first empirical psychologist could also be viewed—at least by some—as an early champion of functionalism. Defined as the view that mental states are constituted solely in relation to their functional role in relation to other mental states, sensory inputs, and behavioral outputs (Putnam 1994), functionalism has been roundly criticized for allowing cognitive scientists (at least in the 1970s and 1980s) to disregard the biological instantiation of mental processes. How could the first physiological psychologist be taken for a functionalist?

These debates probe the distinction between formal and material causes in Aristotle's psychology and its implications for the relation between mind and physical embodiment. Because Aristotle often explains the formal cause of a substance by reference to its function (*De Anima* II 1, 412b 10–24), the functionalist reading of the mind-brain relation claims that formal causes do all the work, such that the physical embodiment or material cause of a thinking substance is incidental. If this were Aristotle's view, perhaps he would be akin to a functionalist. But it is questionable that the formal cause can be explained solely in functional terms. In any case, for the purposes of orienting the following collection of essays, we wish to point out an alternative reading, namely, that Aristotle saw something in the materiality of living things that is just as essential to life and mind as their formal properties. The upshot of this insight is an approach that eschews functionalism by its attention to the details of embodiment, and eschews reductionism by its attention to the emergent properties of the organism as a dynamic, purposive whole. Thus, life and subjectivity cannot be reduced to their underlying materiality, but they are nevertheless wedded to their materiality in ways that non-living systems are not.

The contributions of this volume have been written in this spirit. Through a variety of perspectives and traditions, this book seeks to illuminate possible contributions of philosophy to non-reductive forms of neuroscientific research and thereby to promote a rich form of interdisciplinary exchange. As a whole, the book questions the naïve assumption that the language and concepts of philosophy will eventually be superseded by those of neuroscience, but more importantly, it shows the continuing value of philosophical thought. Each of the contributions addresses one or more aspects of subjectivity in relation to science, including (1) the meaning and scope of naturalism and the place of consciousness in nature, (2) the relation between intentionality, teleology, and causality, (3) the nature of life and its relation to mind, and (4) the role of value in mind and nature. In addressing these issues, the contributions aim to show how philosophy might contribute to real explanatory progress in science while remaining faithful to the full complexity of the phenomena of life and mind. At the same time, the volume displays a considerable width of philosophical support for non-reductive neuroscience, whether by means of hylomorphism, the notion of *enérgeia*, teleology, dynamical systems, enactivism, value theory or ethics. The essays present views from the analytic, phenomenological, and pragmatic traditions of philosophy, all of which have developed their own constructive-critical approach to the problems of reductive science.

This volume brings together the work of scholars that belong to or are in contact with an interdisciplinary research group headquartered at the University of Navarra (Spain). Beginning in 2011, many of the contributors have participated together in conferences and intensive seminars. In selecting the essays, we have tried to strike a balance between accomplished figures and rising young scholars. All of the contributors share the conviction that the way in which the central issues of subjectivity have been approached by science is in need of fundamental change, and that interdisciplinary cooperation is necessary to bring this change about. For this reason, instead of pressing for a particular line of thought, the volume presents a harmony of compatible yet diverse viewpoints.

We now describe the contents of the book chapters and specify their links with the central issues outlined above.

Sturma's chapter is principally concerned with the first issue, the meaning of naturalism and the place of consciousness in nature. He observes that the remarkable success of neuroscience in recent decades has left a host of philosophical problems unresolved, and principal among these is the so-called "Hard Problem" of consciousness. Sturma traces the inadequacy of standard neuroscientific accounts of consciousness to the widespread embrace of an eliminativist form of naturalism that views nature as a closed system and the language of physics as its ultimate semantics. In its place, Sturma recommends a return to the rich variety of non-reductive naturalistic options that are available in philosophy, as exemplified by Aristotle, Leibniz, Spinoza, Schelling, and Hegel, as well as Wittgenstein, Feigl, Sellars, Strawson, among others in the last century. In particular, Feigl, one of the founders of the Vienna Circle, defended an integrative naturalism for which justification must work across disciplines. Moreover, within this integrative naturalism, irreducible features of subjectivity must guide the analysis of issues like personal identity and consciousness. To be successful, this analysis must integrate the data of different perspectives and look at the individual as a psychophysical unity with a multiplicity of equally real dimensions.

Klima's chapter places the same issue into a broader historical perspective. He argues that reductionism is tied to the fact that scientific inquiry continues to labor under the influence of Cartesian thought. Specifically, the well-known post-Cartesian dilemmas of mind and body are based on a set of assumptions that originate in Descartes' famous "Demon-argument." In his analysis of this argument, Klima uncovers premises that were inherited from late-medieval logic, metaphysics, and theology, where the possibility of "Demon-skepticism" first arose, and contrasts these premises with an earlier tradition, exemplified by Thomas Aquinas, which did not allow the conceptual possibility of a "Demon-scenario." Klima argues that the contrastive analysis of these different conceptual frameworks allows us to pinpoint precisely where we need to revise our own conceptual framework in order to rid ourselves of the premises that undergird the mind-body dilemmas that continue to plague the sciences of mind and brain.

Jaworski's chapter presents the hylomorphic notion of structure as a conceptual resource for developing a naturalistic understanding of mind that is non-reductive and non-physicalist but also non-emergentist and non-mysterious. Hylomorphic structures, on this view, are powers of material integration and configuration that belong essentially to individuals as such—to integrate and configure materials in a certain way is what it means to be an individual. Moreover, individual-making structures confer powers on the materials that they configure, powers that these materials do not have on their own. In particular, living individuals are distinguished by their repertoires of "activity-making structures": powers to coordinate parts so as to perform certain tasks or functions. Jaworksi develops the notion of hylomorphic structure so as to take up a subtle position on the natural embodiment of mental properties: on the one hand, as structures, mental powers depend on the powers of the body parts or materials that they configure; on the other hand, body parts alone cannot give rise to mental powers. Thus Jaworski uses hylomorphism to deny both reductive physicalism and bottom-up emergentism.

Buchheim's chapter also makes use of Aristotle's hylomorphism, in this case to articulate the difference between physiological and psychological states. Whereas bodily changes can be attributed to physiological phenomena, changes to the psychological states of the individual are embedded in what he calls a "vital context." In particular, he argues that events that shape the individual's vital context are not spatiotemporally distinguishable in the manner of physical changes, so that psychological and physical changes do not map onto one another. Accordingly, while there is a close relationship between physical and psychological phenomena within the same substance, this relationship cannot be described as a one-to-one correspondence between isomorphic structures. Life episodes turn out to be richer and more multifaceted than somatic states, and this lack of isomorphism helps to explain the causal efficacy of the former. For example, recent experiments on the behavior of monkeys show that anger and disappointment can drive neuronal selection toward better performance at certain tasks. Based on this experiment, Buchheim concludes that, *pace* supervenience theory, life episodes are not causally inert epiphonema. For Buchheim, therefore, non-reductive causal analysis of the "vital context" is essential for the work of neuroscience.

García-Valdecasas' chapter contrasts the principle of causal closure, which has been defended in the context of physicalism, with an Aristotelian view of nature. The principle of causal closure is also known as "the canonical argument for physicalism" and has helped reductionist physicalism to thrive on a number of fronts. The principle of causal closure holds that the perspective of physics is the ultimate arbiter of causality and that its causal analysis is complete for all explanatory purposes. However, reality presents us with many complex phenomena that resist characterization by physics alone. Among these are teleological phenomena, or end-directed behavior. By the close observation of natural regularities, Aristotle understood that the prevalence and sophistication of these regularities cannot be solely attributed to efficient mechanisms. Neither the difficulty of defining the ends of natural substances nor the fact that such ends have been associated historically with "backwards causality" suffice, in García-Valdecasas' view, to rule out the necessity of final causes. For one thing, without reference to final causes, biology fails to distinguish living from non-living processes and, as a result, to identify information that proves crucial to any biological account of a living organism. In this way, it is not that the biologist rejects final causes; rather, she implicitly assumes their existence as a working hypothesis that ultimately illuminates her findings, even if this hypothesis does not feature in her account as such. The recognition of ends does not prevent scientific progress, nor does it get in its way. Rather, their recognition usually marks out its success. Therefore a careful understanding of life requires a theory that can capture and identify ends qua ends. In García-Valdecasas' view and that of some of the leading interpreters of Aristotle, ends are irreducible, built-in, natural dimensions of living processes.

Is it possible to reconcile an objective, scientific description of the body with the subjective experience of being or having a body? Murillo's chapter argues that because an objective account understands subjectivity as a function of the biological organism, it fails to explain the complex and often paradoxical relation between subjectivity and corporeity. Contemporary philosophers such as Plessner, Heidegger and Jonas have pointed out the importance of the concept of life for understanding subjectivity. Moreover, through their analysis of life, they confirm Kant's insight into the close relationship between subjectivity and temporality. It is no mystery that many attempts to reconcile subjectivity and the body fail to capture movement and time, as they often rely on definitions of living systems for which time is always extrinsic. A different and more successful strategy is to invoke the Aristotelian concept of *enérgeia*, or perfect activity, to describe the distinctive character of living processes. This notion also illuminates the role of consciousness, which at face value appears to be out of time, thus posing an even harder problem. For Aristotle, consciousness depends on a vital activity: the activity of the intellect (*nous*). Thus Murillo integrates the discoveries of contemporary biology with key Aristotelian concepts like *enérgeia* and *nous* to develop a concept of the living body as a center of coordinated vital activities, including the generation of consciousness, presence and temporality, and to grasp their characteristic temporality as synchrony. In this way we can envisage a new framework for understanding the body as more than a mere thing and a means of reconciling the atemporality of conscious experience with the instability and exteriority of our body.

Stapleton and Froese's chapter brings together phenomenology and enactivism to provide a scientifically viable account of subjectivity. Drawing from the phenomenology of embodiment developed by Merleau-Ponty, Jonas and others, enactivism is an emerging form of cognitive science that operates within the "fundamental circularity" of subjective experience and scientific explanation. In this chapter, the authors present an enactive conception of agency, which, in contrast to current mainstream theories of agency, is deeply and strongly embodied. They argue that anything that ought to be considered a genuine agent is a biologically embodied (even if distributed) subject, and that this embodiment must be affectively lived in a way that is interoceptively accessible to the subject. These co-dependent biological foundations and phenomenological constraints help to distinguish embodied multi-agent systems (e.g. multicellular organisms), as investigated by physiology, from the social multi-agent systems that are the topic of ethology (e.g. animal herds) and sociology (e.g. societies). Although both kinds of multi-agent system have collective dynamics at the system level and are composed of several agents in their own right, only the former embody intentional agency and subjective feeling at the system level as a whole.

Bower's chapter explores the contributions of the European phenomenological tradition to cognitive science, especially the "embodied and enactive" form of cognitive science that has recently emerged as a robust alternative to the traditional "computational-representational" view. Specifically, Bower examines various ways of describing the world-directedness of personal experience within the phenomenological tradition in an effort to distinguish those perspectives that are best suited to the non-representationalist agenda of "radical" embodied cognitive science. Bower finds that the works of Husserl and Heidegger are susceptible to representationalist readings, while the works of Levinas and Merleau-Ponty lend themselves to a fully non-representational reading of the cognitive aspects of personal experience. Yet the point is not just to show how experience can be purged of representations; more importantly, Bower suggests that the phenomenological analysis of the cognitive aspects of experience indicates why representations are ill-suited to the very cognitive functions they are invoked to explain.

Barrett's chapter examines the role of valuation in mind and natural causation. Within cognitive science, most accounts of value define it as the product of a special subsystem or structure which is extrinsic to basic mental functions of acting, perceiving, and knowing. Not only do such attempts fail to account for our experience of value, but they also fail to recognize the fundamental role of value in all mentality and the corresponding need for value to be grounded in the world. In search of a better account, Barrett's essay explores a naturalistic but non-reductive approach to value based on the American traditions of pragmatism and process philosophy. Rather than attempt to objectify value by denuding it of subjective character, this approach seeks to formulate a general notion of valuation that extends certain value-related characteristics of subjectivity—such as selectivity and differential importance—to all natural processes. The essay also explores the possibility that the ubiquitous phenomenon of collective dynamics, now the focus of several non-reductive research programs in cognitive science, exhibits these very same value-related characteristics.

The final chapter by Cottingham also investigates the place of value in nature, focusing especially on evolutionary accounts of moral values. In Cottingham's view, most modern conceptions of morality are strongly influenced by the kind of historical and genealogical approach that emerged in the nineteenth century. Among the main representatives of this approach is Charles Darwin, who in his *Descent of Man* presents morality as a product of man's struggle for survival. Another influential deflationary account of morality is Mill's utilitarianism, for which the deliverances of conscience are merely psychological events or subjective feelings. Mill portrays normativity as a kind of painful feeling, linked to the violation of duty, which works within the subject as an internal sanction. But a sanction of this kind is just a causal motivator that carries no real moral justificatory force. Such views lead eventually to the radical position of Nietzsche as presented in his *The Genealogy of Morals*: once we realize that ethics has a genealogy, that is, that it is the product of a contingent history, we can transcend this history and the authoritative power of conscience. However, arguments of this kind are hard to reconcile with intrinsic features of morality such as the objective, universal, necessary and normative character of moral values. This paper explores to what extent contingency jeopardizes naturalistic accounts of morality and other contemporary views that seek to reconcile the features of morality with those of the natural world.

References

Barlow, H.B. 1972. Single units and sensation: A neuron doctrine for perceptual psychology? *Perception* 1: 371–394.

Bennett, M.R., and P.M.S. Hacker. 2003. *Philosophical foundations of neuroscience*. Oxford: Blackwell.

Bickle, J. 2003. *Philosophy and neuroscience: A ruthlessly reductive account*. Dordrecht: Kluwer Academic Publishers.

Bullock, T.H., M.V.L. Bennett, D. Johnston, R. Josephson, E. Marder, and R.D. Fields. 2005. The neuron doctrine, redux. *Science* 310: 791–793.

Chirimuuta, M., and I. Gold. 2009. The embedded neuron, the enactive field? In *The Oxford handbook of philosophy and neuroscience*, ed. J. Bickle. Oxford: Oxford University Press.

Choudhury, S., and J. Slaby. 2012. *Critical neuroscience: A handbook for the social and cultural contexts of neuroscience*. Oxford: Blackwell.

Dewey, J. 1972. The reflex arc concept in psychology. In *The early works of John Dewey 1882–1898, Volume 5: 1895–1898*, ed. J. Boydston. Carbondale: Southern Illinois University Press (Original work published 1896).

Freeman, W.J. 1999. *How brains make up their minds*. New York: Colombia University Press.

Gallagher, S. 2006. *How the body shapes the mind*. Oxford: Oxford University Press.

Guillery, R.W. 2005. Observations of synaptic structures: Origins of the neuron doctrine and its current status. *Philosophical Transactions of the Royal Society B* 360: 1281–1307.

Hubel, D.H., and T.N. Wiesel. 1959. Receptive fields of single neurons in the cat's striate cortex. *Journal of Physiology* 148: 574–591.

Juarrero, A. 1999. *Dynamics in action: Intentional behavior as a complex system*. Cambridge, MA: MIT Press.

Kauffman, S.A. 2000. *Investigations*. Oxford: Oxford University Press.

Keeley, B.L. 2009. The role of neurobiology in differentiating the senses. In *The Oxford handbook of philosophy and neuroscience*, ed. J. Bickle. Oxford: Oxford University Press.

Kelso, J.A.S. 1995. *Dynamic patterns: The self-organization of brain and behavior*. Cambridge, MA: MIT Press.

Kuffler, S.W. 1953. Discharge patterns and functional organization of mammalian retina. *Journal of Neurophysiology* 16: 37–68.

Lakatos, I. 1970. Falsification and the methodology of scientific research programmes. In *Criticism and the growth of knowledge*, ed. I. Lakatos and A. Musgrave. Cambridge: Cambridge University Press.

Langton, R. 2000. The musical, the magical, and the mathematical soul. In *History of the mind-body problem*, ed. T. Crane and S. Patterson. New York: Routledge.

Laughlin, R.B. 2005. *A different universe: Reinventing physics from the bottom down*. New York: Basic Books.

Murphy, N. 2006. *Bodies and souls, or spirited bodies?* Cambridge: Cambridge University Press.

Noë, A. 2009. *Out of our heads: Why you are not your brain, and other lessons from the biology of consciousness*. New York: Hill and Wang.
Northoff, R. 2013. *Unlocking the brain: Volume 1: Coding*. Oxford: Oxford University Press.
Northoff, R. 2014. *Minding the brain: A guide to philosophy and neuroscience*. New York: Palgrave Macmillan.
Nussbaum, M.C., and A.O. Rorty. 1992. *Essays on Aristotle's 'De Anima'*. Oxford: Oxford University Press.
Putnam, H. 1994. Putnam, Hilary. In *A Companion to the Philosophy of Mind*, ed. S. Guttenplan, 507–513. Cambridge: Blackwell.
Robinson, D.N. 1976. *An intellectual history of psychology*. New York: Macmillan.
Rosen, R. 1991. *Life itself: A comprehensive inquiry into the nature, origin, and fabrication of life*. New York: Columbia University Press.
Sheets-Johnstone, M. 2011. *The primacy of movement, expanded*, 2nd ed. Amsterdam: John Benjamins Publishing.
Smolin, L. 2006. *The trouble with physics: The rise of string theory, the fall of a science, and what comes next*. New York: Mariner.
Thomson, E. 2007. *Mind in life: Biology, phenomenology, and the sciences of mind*. Cambridge: Belknap.
Varela, F. 2000. *El fenómeno de la vida*. Santiago de Chile: Editorial Dolmen.
Weiss, P.A. 1973. *The science of life: The system of living—A system for life*. New York: Futura.
Wittgenstein, L. 1958. *The blue and brown books*. Oxford: Blackwell.
Wolfram, S. 2002. *A new kind of science*. Champaign: Wolfram Media.
Zahavi, D. 2005. *Subjectivity and selfhood: investigating the first-person perspective*. Cambridge, MA: MIT Press.

Chapter 2
Self-Consciousness, Personal Identity, and the Challenge of Neuroscience

Dieter Sturma

2.1 The Challenge of Neuroscience

In recent decades we have heard many reports of breakthroughs in neuroscientific research and the development of new therapies. It is often assumed that the evident success of neuroscience has a far-reaching impact on many other scientific disciplines, but above all on philosophy. For example, it has been claimed that experiments in neuroscience show that persons have no free will. According to these reports, what a person intends and does is the product of micro-mechanisms which are mostly inaccessible to the reflecting or acting person. Others suggest that scientific research into human consciousness is largely complete and that the mind-body problem already has been or will soon be solved by neuroscience—at best, all that remains to be done is a few detailed investigations into the specific functions of the human brain.

The public attention that neuroscience has attracted can be largely attributed to new imaging techniques for brain processes and neuronal networks. By means of these techniques, neuroscience has managed—wittingly or not—to place the human brain at the center of modern human self-understanding. In discussions of the nature of consciousness, for instance, it seems that we have access to brain images which are supposed to represent basic neural processes underlying specific experiences of persons. One can almost speak of the technical enchantment of human self-

D. Sturma (✉)
Department for Philosophy, University of Bonn, Institute of Science and Ethics (IWE), German Reference Centre of Ethics in the Life Sciences (DRZE), Bonner Talweg, 5753113 Bonn, Germany

Institute of Ethics in the Neurosciences (INM-8), Research Centre Jülich, 52425 Jülich, Germany
e-mail: dieter.sturma@uni-bonn.de

M. García-Valdecasas et al. (eds.), *Biology and Subjectivity*, Historical-Analytical Studies on Nature, Mind and Action 2,
DOI 10.1007/978-3-319-30502-8_2

understanding through neuroimaging. However, it remains dubious whether the more or less popularizing ways in which the media have chosen to introduce the findings of neuroscience to non-expert audiences provide an adequate representation of the state of the art.

The challenge of neuroscience does not stem from its specific scientific approach. With respect to its method and subject of investigation, there is no substantive conflict between the manifest image of man in everyday life and the naturalism of neurobiology and other neuroscientific disciplines. Both are indispensable elements in the project of enlightenment that seeks to combine scientific knowledge with the rational justification of social norms. Neuroscience as such cannot be termed antihumanistic in the sense of undermining our self-understanding of how to lead the life of a person. Indeed, neuroscientific theories should not be confused with eliminativist strategies: contemporary eliminativism is a variant of an older, radical *philosophical* position[1] which, in the guise of a neuroscientific theory, seeks to deprive persons of epistemic and moral attitudes. According to this view, humans are thoroughly contextualized beings who are entirely bound by environmental conditions and occurrences.

The eliminativist approach seems to be justified by the way in which persons are treated like objects in neuroscientific experiments for technical and methodological reasons. But even in these cases, persons are never mere objects. Independently of ethical considerations, persons remain epistemic and practical subjects insofar as they act and behave *like* persons whenever they experience something—apart from cases of severe impairment. Moreover, in many experiments neuroscientists must interact with the patient to discover the correlations between neural mechanisms and states of mind. These facts are either ignored or underestimated in eliminativist approaches, which attempt to sidestep the realm of human behavior and its attendant epistemic and practical attitudes and to characterize persons only with the attributes of objects. One could even say that in some neurophilosophical treatments persons are deliberately converted into things (see Wegner 2002, ch. 4.). It would, of course, be wrong to accuse neuroscience *in toto* of transforming the human subject into a mere object. The vast majority of neuroscientific research is oriented toward therapeutic ends for the benefit of persons, and thus has no inclination to adopt eliminativist or instrumentalist scenarios.

Still, even if eliminativist or instrumentalist views are rejected, it cannot be denied that neuroscience does not leave our self-understanding untouched. It is difficult, however, to ascertain how deep this impact goes. Because progress in the understanding of the neural micro-mechanisms of the brain is rarely linked directly to an understanding of the *mind*, it cannot be expected that genuine problems of the philosophy of mind can be solved by using neuroscientific methods alone. A good example of this difficulty is the phenomenon of self-awareness. The methods of neuroscience have not yet provided direct access to the phenomenon of self-awareness and are therefore unable to offer an adequate explanation of this undeniable feature of human life.

---

[1] The paradigm of the eliminativist position is French Materialism, cf. La Mettrie 1960; d'Holbach 1990.

While philosophy attempts to develop theories about psychophysical relations, neuroscience is content to explore the brain as the physical basis of mental states or phenomena. By means of non-invasive neuroimaging techniques, scientists are able to observe patterns of brain activity underlying specific mental states. In this respect, neuroscience has delivered significant new findings in the last few decades, especially with regard to the biological mechanisms behind emotional and cognitive states.

Nevertheless, these neuroscientific advances have left the psychophysical problem unresolved. We understand neither the nature of the relationship between neural mechanisms and intentional states of consciousness nor how intentional consciousness or self-consciousness can be a result of neural processes. A neuroscientific explanation of the complexity of human consciousness and its phenomenal content seems beyond the reach of current scientific methods. The conceptual schemes used in subjective experience, on the one hand, and in the perspective of the external observer, on the other, differ far too much for easy reconciliation.

Contrary to widespread assumptions, the challenge of neuroscience does not lie in the discovery of facts that demonstrate the validity of neurological determinism or neurological fatalism. Facts do not dictate how they are to be interpreted theoretically, and popular phrases such as *We are nothing more than a pack of neurons!*, *Neurons determine who we are!*, or *Our brain always decides for us in advance!* are merely metaphysical hypotheses dressed up in empirical guise (see Crick 1995; Dennett 1981; Wegner 2002).

Thus, while it appears as though neuroscience and philosophy of mind are dealing with the same subject, they use in fact very different methods to approach mind and brain. These methodological differences are a heavy burden for those who wish to establish greater interdisciplinary collaboration between philosophy and neuroscience. Moreover, it does not seem possible to find connections between the workings of neuronal processes and expressions in the space of reasons. In the perspective of the perceiving, experiencing, and acting person, the everyday world and brain research do not converge. We are always either on the side of self-consciousness and experience, or on the side of neuroscience, studying neural processes from the outside.[2]

2.2 Naturalism

While traditional philosophy of mind mostly emphasizes the uniqueness of self-consciousness, eliminativist approaches of naturalism treat the phenomenon of self-reference mainly as a source of disturbance and irritation. They do not subscribe to the traditional doctrines that self-consciousness is infallible, different from all other states of consciousness, and a state of epistemic privacy and self-familiarity that cannot be derived from other mental states (see Sturma 1985, 2005). Ambitious

[2] The thought experiment of the enlarged brain and the mill by Leibniz exemplifies this problem convincingly: see Leibniz 1704, §17.

efforts to develop a conception of philosophy of mind without self-consciousness have been made time and again since the golden age of positivism and the linguistic turn. Approaches such as behaviourism, physicalism, eliminative materialism, functionalism, and evolutionary epistemology share a common theme: they all react to the phenomenon of self-consciousness with reductionist or even eliminativist methods that basically aim to do away with the specific features of self-consciousness. Each time there were high hopes that the problem of self-reference could be solved for good. The constant succession of eliminativist efforts indicates, however, that these hopes have not been fulfilled. In this situation, it does not seem far-fetched to turn to neuroscience, but like all the other disciplines of the natural sciences, it is, for methodological reasons, ill-equipped to deal with the phenomenon of self-awareness.

In essence, every scientific discipline concentrates on identifying rules and nomologies that explain events and occurrences in the world. For this reason, the natural sciences as well as the humanities have implicitly or explicitly supplied numerous models of the dependencies in the lives of persons. These models presuppose a closed universe of regularities such as those described paradigmatically by modern physics. According to such models, the lives of persons, like the existence of any other spatio-temporal object, are tied into a network of causes—physical, biological or social.

Since the time of Descartes, the natural sciences have left a deep impression on philosophy. Many contemporary philosophers of mind take the authority of natural sciences for granted and search for epistemological reasons to exclude self-referential attitudes and states of persons from the start. Methodologically, they orient themselves exclusively towards the objective perspective of the external observer. These approaches bind their conceptual fate to scientific eliminativism.

Scientific eliminativism is a radicalized version of naturalism. While naturalistic approaches presuppose the ontological unity of reality and the formal unity of the sciences and therefore assume that nature, including intelligent life, is a closed system, scientific eliminativism takes the language of natural science as the ultimate semantics for describing what there is.[3] This approach excludes the most important realm of the philosophy of mind: the epistemic, emotive and moral states of self-conscious persons.[4] A number of properties of experiential states can be addressed in part by scientific theories, but they cannot be separated completely from self-reference.

Eliminativist strategies aim at a simple understanding of the world according to the one-dimensional workings of the realm of causes. One has good reasons to doubt that this understanding is an adequate reaction to the complex reality in which

[3] Scientistic physicalism has a number of problems. They are characterized by ontological generalisations, a very narrow use of deductive methods of explanations, and epistemological vagueness. The internal difficulties of the physicalist approach cannot be further pursued here. For a critique of physicalism see Sturma 1985, 1997, ch. 2; Chalmers 1996, ch. 2; Bennett and Hacker 2003.

[4] The concept of the person manifests itself in theory as well as in everyday life. It can be approached by way of concepts like self-reference, intentionality, reasons for action, life plan, recognition and respect. The semantic core of the concept of a person can be summarized by saying that "person" refers to a being who lives in a social space under the conditions of possible self-consciousness and possible autonomy. Cf. Sturma 1997, 2007.

we live. Moreover, even if naturalistic principles are accepted, it is debatable whether first-person perspectives can be excluded. The history of philosophy has provided other alternatives. We find examples of a non-reductionst naturalism in almost every period of the history of philosophy; this applies to Aristotle, Stoicism, Leibniz, Spinoza, Rousseau, Kant, Schelling, Hegel as well as to Wittgenstein, Feigl, Sellars, Strawson, Hampshire, Hurley, and McDowell—to name but a few. What these various thinkers have in common is that they do not succumb to the temptation of premature eliminations and they all work with a complex model of reality, which has monistic features in its ontology and dualistic features in its epistemology. They follow a method that avoids the extremes of scientific eliminativism without falling back on dualist positions of mind and body.

A good example of non-eliminativist naturalism is the position of Herbert Feigl. As one of the founders of the Vienna Circle, he subscribed to the paradigm of scientific realism and to a narrow form of naturalism, but he could not be convinced to treat the psychophysical problem as a pseudo-problem (cf. Feigl 1967; cf. Sturma 1998). In his various approaches to linguistic and epistemological issues, he incorporated dual aspects into an overall monist framework. While he conceded that "there is something which purely physical theory does not and cannot account for," (Feigl 1967, p. 109.) he did not mean that we can simply fall back on traditional forms of anti-reductionism or dualism. Feigl had no confidence in traditional positions when it came to explaining mental causation. Instead, he assumed that the place of mind in nature can only be accounted for within the framework of a naturalism that allows for correlations between phenomenal contents and causal mechanisms. According to Feigl, the philosophy of mind has to work with co-references between a subjective point of view and knowledge by description. This co-reference has to be seen as "a characteristic of the basic nature of our world." (Feigl 1967, pp. 110–11).

Feigl's general approach can be labelled integrative naturalism. Integrative naturalism holds onto the ontological unity of reality and the unity of the sciences in the sense that justifiable assertions in different disciplines cannot contradict one another. Scientific justification works across disciplines. But an integrative naturalism rejects eliminativism and the primacy of a particular scientific position. It distinguishes between the space of reasons and the realm of causes, and grants both of them a non-eliminable methodological status. Integrative naturalism emphasizes the crucial role of semantics and epistemology in theoretical models, and, contrary to radical physicalism, it insists that there are no brute facts that should predefine the scope of naturalism.

2.3 Self-Consciousness

Descartes revealed that except for immediate self-awareness and analytical truths, all propositions or propositional attitudes are in principle open to error. Neither tradition nor sense perception nor linguistic reference is immune to doubt. But when I express openly or tacitly the sentence, "I exist," it is impossible for me to be wrong. The linguistic expression of my existence cannot be doubted for the duration of its expression. The sentence, "I do not exist," is self-defeating, as it presupposes the very existence that it denies.[5]

The innovative elements of Descartes' approach stem from his discovery of the principle of inwardness. According to this principle, self-referential consciousness provides the basis of knowledge. The core of the principle of inwardness is the *existo* argument in his *Second Meditation*. This passage does not contain the notorious phrase *cogito ergo sum*. The evidence for the *existo* argument is already accessible from the perspective of everyday experience. The reasons for this evidence, however, can only be revealed in epistemological reflection, and it is likely that even Descartes himself was unaware of these reasons.

The certainty of the *existo* argument depends on the semantic context of indexicals and self-referential expressions: the assertion must be made in the first-person singular, present tense, indicative, and active form. The self-referential expression of existence is specified by the expression "I." But the indicating function of the expression "I" should not be confused with the "I" as a placeholder for a name or a metaphysical concept. The transformation of the indicator "I" into a substantive or a term that stands for a particular is a fallacy.

With the principle of inwardness, Descartes has left us a complex heritage. An essential part of this heritage consists in the reification of the self—i.e. in the assumption that the term "I" refers to a thinking thing. This assumption makes traditional philosophy of mind seem unacceptable to many contemporary philosophers, but the reification of the self is only one part of the Cartesian heritage. Descartes' legacy also includes his discovery of the systematic function of the *existo* or *cogito* and the immunity of self-consciousness to eliminative strategies. The Cartesian heritage is thus a complex matter, and one should shy away from hasty rejections.

What I refer to in self-consciousness is not my body or any other empirically identifiable state, and yet self-consciousness must contain a referent, or at least an intentional correlate, which makes the possessive situation of self-consciousness possible. Self-awareness stands out by being self-familiar and infallible—an infallibility that derives from the fact that in self-consciousness there is no reference to a physically identifiable object. Nevertheless, self-consciousness must imply some form of reference through which it can express itself in time. Without any referential structure at all, it would be an empty thought and would be nothing to me.

However, the reference of self-consciousness cannot be understood as purely self-referential. Rather, a person attains self-awareness by implicitly referring to something different from herself. The remarkable quality of self-consciousness consists in the fact that reference arises through an act of self-reference that in itself

[5] See Descartes 1964, II. 3; cf. Ayer 1956, pp. 44–52. Descartes operates in this passage with the formula "I exist." Yet, the certainty of self-consciousness expresses itself also in the formula "I think." The "I think" (*cogito*) covers all forms of mental activities (*cogitationes*). That is why a sentence like "I *believe* there is a unicorn" is true as long as it is used authentically—regardless of the fact that unicorns do not exist in the physical world.

is referentially differentiated. Accordingly, a person can become aware of herself only by attributing to herself states that refer—in whatever way—to her position in the world of possible experiences. This position is the ontological condition of self-reference, and only because of this is the perspective of self-consciousness a perspective in the world.

Since self-reference is the necessary condition for all epistemic states of human consciousness, it cannot refer to a specific object. The difference between the identity of self-consciousness over time and personal identity in all self-ascriptions over time indicates the dividing line between the semantics of the expressions *self-reference* and *person*, as well as between epistemology and philosophy of the person in a narrower sense (see Kant 1781/1787, B 409). Whereas identity of self-consciousness expresses itself in the self-referential structures of mental acts, further epistemic, moral, and social requirements have to be met in the case of personal identity.

The difference between self-reference and self-identification is the outcome of epistemic relations and does not directly express different ontological structures. Identity of self-consciousness and identity of the person cannot stand side by side without being related. There has to be a structural similarity, or an isomorphism, between identity of self-consciousness and identity of the person, otherwise the fact that persons are at least sometimes aware of themselves in the world of possible experiences would not itself be possible. In the case of the lives of persons, the extensions of the conditions of unity vary: physical unity has a wider scope than unity of consciousness, and psychological identity has a wider scope than practical or moral continuity over time. Not every case of physical identity is a case of unity of consciousness, but every case of unity of consciousness implies physical unity. The same is true for the relation of psychological identity and practical continuity over time.

2.4 The Identity of a Person Over Time

From the subjective perspective of a person, self-awareness is the key to understanding her own identity, but it is not sufficient for constituting it. The theoretical framework has to be expanded beyond the concept of self-consciousness into the realm of physical constituents. Identity of a person over time is a complicated phenomenon, which has to be distinguished from physical or psychological identity in the narrow sense. The identity of a person over time is an occurrence which expresses itself internally *and* externally—through the person's mind as well as through her body. Personal identity is embodied and in this respect at least partially accessible from the point of view of an external observer.

The identity of self-consciousness is, from the subjective perspective, infallible and self-sufficient, but it must comply with external conditions. Due to the epistemic asymmetry of the perspectives of the first, second, and third person, the identity of a person unfolds in a twofold way. Subjectively, self-awareness manifests itself as self-consciousness over time (cf. Kant 1781/1787, B 412 FN). In order to gain an understanding of the continuity of a person over time in its entirety, a person has to refer to psychological and physical constituents of her continuity in space and time.

The epistemic difference between the identity of self-consciousness and the identity of the person amounts to a specific connection between the 'simple view' and the 'complex view.' The simple view understands self-consciousness and personal identity as facts in the sense of intrinsic non-derivative qualities (cf. Chisholm 1976, 1981; Swinburne 1986). From this approach, it is possible to consider the possibility of the persistence of personal existence after the disintegration of the body. In contrast, the complex view assumes that self-consciousness and personal identity have to be understood on the basis of physical or psychological relations and continuities (cf. Williams 1978, 1984; Wiggins 1967; Lewis 1970; Parfit 1984). Whereas the simple view assumes that the subjective perspective of experience and the identity of the person are irreducible, the complex view starts from the assertion that consciousness and identity of the person are composed of physical or psychological relations or are at least deducible from them.

Contrary to the received opinion, there is no rigid opposition between the simple view and the complex view. Instead, it is possible to relate them to each other as methodological elements fulfilling different tasks in explaining the life of a person *as* a person over time. Persons live their lives as a self-referential psychophysical unity over time, and one can work with the irreducibility and the epistemic inaccessibility of self-consciousness without denying that only persons who exist in space and time may be subjects of consciousness and experience.

The identity of self-consciousness and the presence in space and time function as a principle of individuation. This separates the self-aware person from other persons and marks the spatio-temporal boundaries of her existence. To this extent, the epistemic asymmetries of self-awareness and personal identity over time replace the traditional *principium individuationis*.[6] Following the new version of the principle, what it is like to be a person cannot be ascribed to functions or to properties of objects.

Self-consciousness is the starting point for questioning the identity of the person (cf. F. W. J. Schelling 1856–1861, 353 et seq.). Ontologically, the problem of identity is reduced to the forms of existence. Existence simply means being present at a certain time and a specific place: "[A]lthough time and space are equally essential for the identification of the thing, the existence of the thing is essentially temporal; the limits within which it exists are dates, not places. For an extended thing *to be*

[6]The replacement was the innovative move of John Locke. He answered questions of the *principium individuationis* with the introduction of the psychological conception of personal identity over time. See Locke 1975.

thus means to be present *somewhere* in space during a *certain* time." (Tugendhat 1975, p. 28). Extensions of the concepts used to designate the forms of existence vary according to the respective ontological, semantic, and methodological contexts.

2.5 Manifestations of the Mind

The epistemic asymmetries which accompany self-consciousness reveal the structurally different determinations in the phenomenon of personal identity. Within her self-relation, a person combines opposing perspectives of reflection. In contrast to the identification of objects of the "outside world," one does not need any particular criteria for ascribing identity to oneself. As long as personal identity can be experienced from a subjective point of view, there is no need for an ontological examination—with the exception of some severe psychopathological cases.

The certainty of the identity of self-consciousness does not tell a person anything about the empirical content of her existence, however. To fully grasp the numerical identity of a person existing in space and time, one needs empirically accessible situations which fall into the realm of causes or the realm of external reflection. Thus, in order to reconstruct the phenomenon of personal identity in its entirety one has to go beyond the phenomenon of self-awareness—but without leaving it behind. Even when extending the scope of reconstruction, self-consciousness retains a crucial role. In the lives of persons, emotional and cognitive states of consciousness always take place under conditions of *possible* self-consciousness; otherwise they would not belong to the continuous episodes of the mind. Persons are not always self-aware and are often driven by unconscious dispositions and impulses, but as long as they are mentally active it is possible for them to attain self-awareness in every state of mind. When a person is self-aware she experiences explicitly that *she* is in this or that situation.

The identity of a person over time always demands the possibility of self-consciousness. Otherwise, the psychological and physical components of conscious life would fall apart.

In spite of the certainty and infallibility of self-awareness, the personal standpoint depends in its subsistence on external conditions. In particular cases of self-understanding or self-knowledge, the person has already left the state of self-certainty and understands herself as an object of actual or possible experience, which is also accessible to other persons.

Both the epistemological as well as the practical perspectives of personal life are constituted by complex interdependencies between self-reference and reference. These interdependencies refer to a complex system of irreducible physical and psychological components which "carry" personal identity through time.

Because of this psychophysical unity, experiences leave traces in the realm of physical causes. Experiences are not elusive. They are events in space and time of which it often can be said—retrospectively, from a subjective as well as an objective perspective—that they belong to their lives. The continuity of consciousness, dispositions, and actions of a person manifest themselves in the psychophysical identity of a person over time.

One can speak, therefore, of the dynamic interdependencies between experiences and neuronal states as of manifestations of the mind. A change in one of the states has an effect on the other type of states as well—although these changes are in most cases epistemically inaccessible. However, a fundamental change of one of the components can have such an impact that personal identity is severely affected. Personal identity as such does not depend on every little alteration of mind and body. It simply depends on a core area of psychophysical unity. Accordingly, personal identity does not necessarily end only with biological death. There can be cases of psychological deterioration that cause a "loss" of personal identity.

The psychophysical unity and the interdependencies of mental and physical states are the foundational blocks of the practical unity of a person over time. A motive for a specific action is not an ephemeral episode (cf. Hampshire 1959). Attitudes and dispositions endure over time and remain stable under conditions of constant alteration. What a person wants or does and how she acts has an impact on her identity. The dispositions and practical attitudes of a person are of fundamental importance for the constitution of personal identity.

Persons live their lives as psychophysical subjects under the condition of possible self-consciousness. Therefore, it is not sufficient to regard persons as individuals to whom both personal and empirical attributes can be ascribed. Instead, the attributes must be envisioned as *internally* connected and interdependent. Theories which try to set bodily identity, psychological identity, and practical unity apart from each other or on top of each other miss the undeniable fact of an internal unity or coherence between mind and body that is crucial for the concept of personal identity.

A person is obviously a persisting biological entity, but this does not mean that personal identity over time can solely be explained by means of biology. Neither the concept of bodily identity nor the concept of narrative or biographical continuity is sufficient to explain the fact that a person endures as a self-aware being.

Looking at the problem of personal identity and considering the advantages and disadvantages of the simple view on the one hand, and the complex view on the other, it becomes clear that the solution of this intricate problem does not lie in simply choosing one of the two alternatives. It is rather a connection between these views that fits the internal connection of self-reference and reference. While the concept of a simple view sees personal identity as non-reducible and epistemologically inaccessible, the complex view regards personal identity as an empirical relation, which is in part accessible. The integrative naturalism that is argued for here takes the irreducibility of self-consciousness for granted while emphasizing that persons can only be subjects of consciousness and experience qua embodied beings.

Physical as well as psychological relations do not consist of simple causal units. It is rather the attitudes and intentions that are expressed in these relations which persist in the life of a person. The function of attitudes and intentions is not merely to start a certain action and then to disappear; instead, they endure in behavioral patterns over time. The inside and outside of personal identity go hand in hand and are only distinguishable in epistemological reflection. On these grounds, personal identity is to be conceived of as a manifestation of the mind that initiates processes in the space of reasons which in turn lead to outcomes and effects in the realm of causes that are not solely under the control of the acting person.

What a person thinks or does leaves traces in her neural system—however marginal. Her mind is neurally incorporated and thus has a physical expression. Although neuroscience has no direct access to self-awareness, it can identify aspects of the manifest mind. However, only an integrative naturalism is in a position to meet the requirements for explaining the astonishing fact of a person's being in the world.

References

Ayer, A.J. 1956. *The problem of knowledge*. London: Macmillan.

Baker, L.R. 2000. *Persons and bodies. A constitution view*. Cambridge: Cambridge University Press.

Bennett, M.R., and P.M.S. Hacker. 2003. *Philosophical foundations of neuroscience*. Oxford: Blackwell.

Bermúdez, J.L. 1998. *The paradox of self-consciousness*. Cambridge, MA: MIT Press.

Cassam, Q. 1994. *Self-knowledge*. Oxford: Oxford University Press.

Castañeda, H.-N. 1999. *The phenomeno-logic of the I. Essays on self-consciousness*. Bloomington: Indiana University Press.

Chalmers, D. 1996. *The conscious mind. In search of a fundamental theory*. Oxford: Oxford University Press.

Chisholm, R.M. 1976. *Person and object: A metaphysical study*. London: Routledge.

Chisholm, R.M. 1981. *The first person: An essay on reference and intentionality*. Minneapolis: University of Minnesota Press.

Crick, F. 1995. *The astonishing hypothesis. The scientific search for the soul*. London: Scribner.

d'Holbach, P.Th. 1990. *Système de la Nature*, ed. J. Boulad-Ayoub. Paris: Librairie-Arthème Fayard.

de La Mettrie, J.O. 1960. L'Homme Machine. In *La Mettrie's L'Homme Machine. A study in the origins of an idea*, ed. A. Vartanian. Princeton: Princeton University Press.

Dennett, D.C. 1981. *Brainstorms. Philosophical essays on mind and psychology*. Brighton: Harvester Press.

Descartes, R. 1964. *Meditationes de Prima Philosophia*. In *Œuvres de Descartes*, VII, ed. Ch. Adam and P. Tannery. Paris: Vrin.

Evans, G. 1982. *The varieties of reference*. Oxford: Clarendon.

Feigl, H. 1967. *The "mental" and the "physical". The essay and a postscript*. Minneapolis: University of Minnesota Press.

Hampshire, S. 1959. *Thought and action*. London: Chatto & Windus.

Hurley, S. 1998. *Consciousness in action*. Cambridge, MA: Harvard University Press.

Kant, I. 1781/1787. 1956. *Kritik der reinen Vernunft*. Hamburg: Meiner.

Leibniz, G.W. 1704. Nouveaux Essais sur l'Entendement humain. In *Philosophische Schriften*, Bd. 5. Hildesheim: Olms 1978.

Lewis, D. 1970. An argument for the identity theory. *Journal of Philosophy* 63: 17–25.

Locke, J. [1694]. 1975. *An essay concerning human understanding*, ed. P.H. Nidditch. Oxford: Clarendon Press.
McDowell, J. 1994. *Mind and world*. Cambridge, MA: Harvard University Press.
Moran, R. 2001. *Authority and estrangement. An essay on self-knowledge*. Princeton: Princeton University Press.
Parfit, D. 1984. *Reasons and persons*. Oxford: Clarendon.
Perry, J. 2002. *Identity, personal identity and the self*. Indianapolis: Hackett Publishing Company.
Schelling, F.W.J. 1856–1861. Fernere Darstellung aus dem System der Philosophie. In *Sämmtliche Werke*, IV, ed. K.F.A. Schelling. Stuttgart: Cotta.
Sellars, W. 1963. *Science, perception and reality*. London: Routledge & Kegan Paul.
Sellars, W. 1974. … this I or he or it (the thing) which thinks …. In *Essays in philosophy and its history*, 62–92. Dordrecht: Reidel.
Sellars, W. [1956]. 1997. *Empiricism and the philosophy of mind*. Cambridge, MA: Harvard University Press.
Shoemaker, S. 1963. *Self-knowledge and self-identity*. Ithaca: Cornell University Press.
Shoemaker, S. 1968. Self-reference and self-awareness. *Journal of Philosophy* 65: 555–567.
Shoemaker, S., and R. Swinburne. 1984. *Personal identity*. Oxford: Blackwell.
Strawson, P.F. 1959. *Individuals. An essay in descriptive metaphysics*. London: Methuen.
Strawson, P.F. 1966. Self, mind and body. In *Freedom and resentment and other essays*. London: Methuen 1974.
Sturma, D. 1985. *Kant über Selbstbewußtsein. Zum Zusammenhang von Erkenntniskritik und Theorie des Selbstbewußtseins*. Hildesheim: Olms.
Sturma, D. 1997. *Philosophie der Person. Die Selbstverhältnisse von Subjektivität und Moralität*. Paderborn: Schöningh.
Sturma, D. 1998. Reductionism in Exile? Herbert Feigl's identity theory and the mind-body problem. *Grazer Philosophische Studien* 54: 71–87.
Sturma, D. 2005. *Philosophie des Geistes*. Leipzig: Reclam.
Sturma, D. 2007. Person as subject. In *Dimensions of personhood (Journal of Consciousness Studies)*, ed. H. Ikäheimo and A. Laitinen. Exeter: Imprint Academic.
Sturma, D. 2008. Die Natur der Freiheit. Integrativer Naturalismus in der theoretischen und praktischen Philosophie. *Philosophisches Jahrbuch* 115, 2: 385–396. Halbband.
Swinburne, R. 1986. *The evolution of the soul*. Oxford: Clarendon.
Taylor, Ch. 1985. *Human agency and language*, Philosophical papers, 1. Cambridge: Cambridge University Press.
Tugendhat, E. 1975. Existence in space and time. *Neue Hefte für Philosophie* 8: 14–33.
Wegner, D.M. 2002. *The illusion of conscious will*. Cambridge, MA: Bradford.
Wiggins, D. 1967. *Identity and spatio-temporal continuity*. Oxford: Blackwell.
Wiggins, D. 1980. *Sameness and substance*. Oxford: Blackwell.
Wiggins, D. 1987. Truth, invention, and the meaning of life. In *Needs, values, truth. Essays in the philosophy of value*. Oxford: Blackwell.
Wilkes, K.V. 1988. *Real people. Personal identity without thought experiments*. Oxford: Clarendon.
Williams, B. 1973. *Problems of the self. Philosophical papers 1956–1972*. Cambridge: Cambridge University Press.
Williams, B. 1978. *Descartes: The project of pure enquiry*. London: Penguin.
Williams, B. 1984. The scientific and the ethical. Royal Institute of Philosophy Lectures 17: 209–228.
Wright, C., B.C. Smith, and C. Macdonald. 1998. *Knowing our own minds*. Oxford: Clarendon.

Chapter 3
Mind vs. Body and Other False Dilemmas of Post-Cartesian Philosophy of Mind

Gyula Klima

3.1 Introduction: What Are False Dilemmas and Why Are They Important?

Since the title of this paper promises a discussion of false dilemmas in the philosophy of mind, we should start the discussion with the notion of a false dilemma. A false dilemma is usually presented by the assumption of a false disjunction. But in the sense intended here, not just any false disjunction expresses a false dilemma. A false dilemma or dichotomy in the stricter sense intended here is a disjunction that is false because of the falsity of a tacitly assumed presupposition, on the basis of which the dilemma would appear to be necessarily true, exhausting all possibilities by mutually exclusive alternatives. A trivial example would be the claim that the present king of France is either bald or he has hair. However, despite possible appearances to the contrary, this cannot be true. On the one hand, if the present King of France has hair, then he exists, but how can he exist, when we know that there is no such a person? On the other hand, if he is bald, then, even if he has no hair, he at least has a head with living skin on it; so he is alive, he exists, whereas, again, we just know that there is no such a person. So, it seems that accepting either horn of the apparently necessarily true dilemma lands us in absurdity. Clearly, the trouble here is caused by the apparent exhaustiveness of the disjunction expressing our dilemma, based on the obviously false presupposition of the existence of its subject. But this is precisely the reason why this dilemma is trivial, namely, because of the obvious falsity of this presupposition.[1] By contrast, the non-trivial, intriguing

[1] For a discussion of treatments of Russell's problem-sentence in medieval logic *without* his theory of descriptions, and also without the false presupposition that reference commits one to existence, see Klima 2001, pp. 197–226.

G. Klima (✉)
Fordham University, New York, NY, USA
e-mail: klima@fordham.edu

M. García-Valdecasas et al. (eds.), *Biology and Subjectivity*, Historical-Analytical Studies on Nature, Mind and Action 2,
DOI 10.1007/978-3-319-30502-8_3

dilemmas of science and philosophy are those whose presuppositions are so deeply ingrained in our actual conceptual framework that they go usually unchecked, so that their tacit, unreflected assumption makes our false dilemmas appear to be exhaustive, and consequently true of necessity. Indeed, such are those dilemmas the anomalous consequences of which prompt the re-checking of our presuppositions, which, upon turning out to be false, may in turn prompt further revisions, and eventually even a complete overhaul of the entire presumed conceptual framework.

One such dilemma concerned, for instance, the nature of light, which on the basis of classical mechanics had to be deemed to consist either in ether waves or of particles. But the anomalous consequences of either alternative (say, on the wave theory there could be no photoelectricity, while on the particle theory there could be no interference or polarization of light) prompted a reconsideration of the basic presumptions of classical mechanics, eventually leading to the "paradigm-shift" to quantum mechanics. To be sure, this brief reminder of the recent history of physics merely serves to illustrate the role of non-trivial false dilemmas in the progress of science. No wonder it is usually upon stumbling over such false dilemmas that scientists tend to turn philosophical. After all, philosophy, being the science of first principles and causes according to Aristotle, is all about checking our otherwise unchecked presumptions, which, as we could see, are precisely the grounds of our false dilemmas.

3.2 A "Catalog" of False Dilemmas of Modern Philosophy of Mind

So, what are the false dilemmas in the modern philosophy of mind I announced in the title? To give the discussion some order, I shall list these moving from "inside out", as it were, starting with the ontological character or essential nature of our subject matter, and ending with dilemmas concerning its characteristic operations. The first dilemma in this order can be formulated as the following disjunctive question: is a human being a mere physical, biological entity or is it perhaps the combination of an organic, physical body and a non-physical entity, a purely spiritual mind or soul that is supposed to survive the death of the physical body?

Of course, this question is only one way to formulate the chief methodological, conceptual dilemma of modern philosophy of mind, namely, that of materialism or *physicalism* vs. *dualism*. To be sure, nobody doubts that viewed in terms of plain common sense we are physical, biological entities, but at the same time there are phenomena in a human life that just do not seem to fit into a simple physicalist framework. On the one hand, on the active side of human life, since moral responsibility seems to go hand in hand with action governed by free choice, we can blame or praise moral agents only for their own actions determined by their choices. However, a consistent physicalist would have to describe human behavior just as any other natural phenomena, namely, in terms of laws of nature, whether those laws be necessary or merely stochastic. But these laws are still deterministic in the broad sense that given certain physical parameters of a natural phenomenon, they determine other physical parameters of the same, even if they may allow some statistical uncertainty. Thus, they seem to leave no room for *autonomous* human action determined (at least in part) by the *free choice* of the agent that itself is not determined by any physical parameters (either necessarily or stochastically), which, however, is the *sine qua non* of moral and legal responsibility. It is along these lines, then, that siding with physicalism may lead to the further specific dilemma concerning human agency, namely, that of *freedom of the will* vs. *causal determinism.* On the other hand, on the cognitive side of human life, there is the issue of phenomenal experience, constituted by the so-called *qualia*, which seem to defy any classification as physical phenomena, given their non-physical character, based on their intersubjective unobservability or, in other words, merely introspective accessibility.

However, if prompted by these considerations one is opting for dualism concerning human nature, then one has to face a number of further false dilemmas concerning the nature, powers and operations of the alleged "ghost in the machine" that supposedly accounts for these phenomena. Indeed, again, moving "from inside out", in a dualist framework one is tempted to regard this alleged non-physical entity as one's true self, as opposed to the biological body it uses, or at any rate to which it is obviously more intimately connected in its operations than to other bodies, yielding the eponymous dilemma of this chapter, namely that of *mind* vs. *body*, as well as its basically equivalent formulations in terms of the contrast between "the mental" and "the physical", or between "my conscious self" inaccessible to others, and the publicly observable biological human being, which on this account would be just a sort of "exoskeleton" of the self, identified as the human person, entrapped in it and operating it with more or less success from within.

But taking this horn of the previous dilemma gives rise to a host of further dilemmas concerning how this operation can possibly take place. Again, on the active side, how on earth can an alleged non-physical entity (lacking precisely those physical parameters, such as mass, momentum, electric charge, magnetic polarity, etc. in terms of which we describe the behavior of physical entities as obeying certain laws of nature) act on a physical entity? After all, even if it may take something that phenomenally appears to us as "willpower" to make certain physical efforts with our own bodies (which are just as ordinary physical, biological bodies as any others), that "willpower" just does not seem to have a physical impact on other bodies, notwithstanding alleged observations of "paranormal" phenomena, which, as such, are definitely not supposed to be the results of the "normal", natural operations of the human mind that would be the concern of "normal" natural science.

On the cognitive side of human life, the dualist scenario seems even trickier. For clearly, the non-physical mind entrapped in and operating the body gets its information about its environment either *directly*, through the direct impact of external bodies on it, or *indirectly*, through its own subjective experiences, the so-called *qualia*, which we might as well call what early modern philosophers after Descartes called them, namely, *ideas*. Of course, this is one way of putting the familiar dilemma of *direct realism* vs. *representationalism*. However, neither alternative seems particularly promising. On the one hand, the direct physical impact of external bodies on the non-physical mind is even less plausible than the impact of the mind's *own* body on it, which is just as problematic as the mind's impact on its body, there being no laws of physics adequately accounting for the interaction between a physical entity and an entity that is non-physical by definition. On the other hand, the mind's getting information about its physical environment (which might as well include its own body) indirectly, through its mental representations or ideas is even more problematic. For on the post-Cartesian representationalist account, we are supposed to infer the presence or absence of external physical entities and their real properties through the presence or absence of their ideas observed by our mind. However, as Thomas Reid correctly pointed out, there is no good inference of this sort, since the idea and the object it represents are obviously distinct items, which means it is always possible to have the one without the other, whence the presence of the one does not entail the presence of the other (cf. Greco 2004, p. 143). Now, perhaps curiously, in a famous passage (AT VII 102; CSM, II 75) Descartes himself explicitly identifies the idea of the sun with the sun itself, as being the single object of the mind attending to it. However, the identity Descartes had in mind clearly had to be *contingent* identity, that is to say, as long as both the idea and its object are present, they are the same, but still, they can become not the same, precisely when either of them persists without the other, as when I close my eyes the visual idea of the sun is gone, while the sun may still be there, or when the sun goes down and I have a vivid dream of a sunny landscape from which in my dream I look up to the dazzling sun, my idea is there, whereas the sun is not (see Klima 2011).

In fact, it is precisely this feature of post-Cartesian theories of mental representation that, as I will soon argue in more detail, allows the apparent possibility of Demon-skepticism, the very starting point of the Cartesian conception of the mind as the thinking conscious self, directly aware of its own ideas, but perhaps nothing else that these ideas appear to represent, apart from one single idea, namely, the idea of this thinking self itself.

However, as our historically better informed colleagues know, the trouble did not start with Descartes. In fact, he inherited the idea of Demon skepticism from late-medieval nominalism, stemming from the work of William Ockham. Indeed, as I will argue, it is Ockham and his followers who are ultimately responsible for the last item in my "catalog" of our false dilemmas in the philosophy of mind and language, namely, the alleged dilemma of *externalism* vs. *internalism* in mental and linguistic representation. Therefore, in "rolling back" the series of false dilemmas laid out thus far, I will proceed in the reverse direction, "from outside in", starting with the issue of Ockham's externalism, as opposed to his predecessors' alleged internalism, as if that were the only possible alternative, which, as I will argue, definitely does not hold for the conception of Thomas Aquinas.

3.3 Ockham's Externalism

Claude Panaccio describes the forms of externalism he identifies in Ockham in the following way (Panaccio 2015):

- *Linguistic externalism* is the thesis that the meaning of the words a speaker utters does not solely depend on the internal state of the speaker at the moment of their utterance, but on certain external factors as well.
- *Mental content externalism* is the thesis that the very content of what an agent thinks does not solely depend on the internal states of the agent, but on certain external factors as well.
- *Epistemic externalism* is the thesis that what an agent believes or knows does not solely depend on the internal states of the agent, but on certain specific external factors as well.

Here I will start the discussion with Ockham's linguistic externalism, however, given the medieval conception of linguistic meaning, what I have to say will have obvious consequences concerning the other varieties described here, especially, mental content externalism. Within the scope of the present discussion, I will deal only with the meaning of what are commonly called "natural kind terms", such as 'man' or 'donkey', but again (with obvious provisos) what I will say can easily be generalized to other linguistic categories as well.[2]

So, how does Ockham characterize the meaning of such a common term? And how and why is that characterization externalist? To answer these questions, we should know in the first place that for Ockham, in line with the medieval Aristotelian tradition in general, a spoken term in itself is just an articulate sound, an utterance or the corresponding written marks or any other easily producible and reproducible entity (say, gestures, etc.) that can acquire meaning through linguistic convention, or as the medievals would have it, by *imposition*. An act of imposition is any socio-linguistic mechanism ranging from ceremonial baptisms to unceremonious slips of tongue or typos (as happened with 'Google', according to one urban legend) that establishes a connection between an utterance and a naturally representative mental act, a human concept. But no matter *how* this connection is established, once it *is* established, the utterance in question becomes a meaningful or significative utterance subordinated to the concept, inheriting, as a result of this conventional imposition, the natural semantic or representative features of this concept. Thus, linguistic meaning is conventional simply because imposition and the resulting subordination

[2] The subsequent discussion of Ockham's externalism and its contrast with what I call Aquinas' "hyper-externalism" partly overlaps with my presentation of basically the same issues in Klima 2015a. However, while the discussion there focused on paradigmatically different medieval conceptions of *semantic content* and their relation to the contemporary contrast between linguistic internalism and externalism, the point of the discussion here is to show how the nominalist conception of *mental content* contributed to the emergence of the modern notion of subjectivity, in stark contrast to an earlier paradigm that avoids the false presuppositions and consequent dilemmas of this modern notion.

of utterances to concepts are conventional. But the conventionally acquired meaning of the utterance is nothing but the natural representational content of the concept to which it is conventionally subordinated.

Looking at the picture so far, it might seem that if this is indeed the general medieval Aristotelian doctrine, then this entire tradition must be inherently internalist: after all, the linguistic meaning of a term seems to be entirely determined by the representational content of the concept to which the term is subordinated, which in turn is nothing but a mental act or mental state of language users; *ergo*, linguistic meaning is entirely determined by internal mental states, which is precisely the definition of linguistic internalism. So, how can Ockham possibly be claimed to be an externalist, once he shares the assumptions of this tradition?

According to Panaccio's intriguing interpretation of Ockham's semantics and cognitive psychology, the representational content of the concepts in question is two-fold: one has to distinguish (1) their objective content, which I will call their *semantic content*, from their (2) built-in *sensory recognition schemes*, which I will call their *phenomenal content*. As Panaccio convincingly argues, these two types of representational content are merely contingently related according to Ockham: the semantic content of a concept is nothing but the first object of a certain natural kind (in particular, of a *species specialissima*, a *most specific species*) that the cognitive subject first encounters as well as all present, past, future and merely possible co-specific individuals represented indifferently by the same mental act, simply because just any of these co-specific individuals could have caused the same act in the same mind, on account of the sameness of their causal powers and the sameness of the representative capacities of the mind representing them. By contrast, the *phenomenal content* of such a concept is the characteristic sensible features of such an object as perceived by the cognitive subject (one may even say, it is the collection of *qualia* or *sensory ideas* this sort of object is naturally apt to evoke in the cognitive subject).

The reason for the divergence of these two types of content is that while the first is determined by the causal laws of nature concerning the natural kind of the object and the cognitive subject, the second is merely determined by whatever the subject happens to perceive in the first object it encounters, or at any rate, the necessarily finite sample of objects of the same kind it has observed, as opposed to the potential infinity of the objects of the same natural kind. For on Ockham's account of concept acquisition, the first sensory encounter with the first instance of a natural kind naturally gives rise to an intuitive singular sensory cognition, as well as a singular, intuitive intellectual cognition of the same object. But what accounts for the singularity of these intuitive acts of cognition is not their distinctive representational content: on the contrary, the representational content of these acts indifferently represents any other sufficiently similar object; in the case of intellectual cognition, any *essentially* similar object, i.e., any other object *of the same natural kind, indifferently represented by this act*. Thus, the singularity of the first intuitive acts is merely due to their actual causal contact with the singular object in question. Therefore, removing the object, or severing the causal connection in any other way, immediately renders the remaining intellectual act an *abstractive act* of cognition, a common concept, indifferently representing all co-specific individuals, which will then constitute its objective, semantic content (See Panaccio 2004: 119–125.).

By contrast, the phenomenal content of the same act is whatever the subject typically perceives of objects of this kind, the typical way objects of this kind would appear to this subject. However, since the perceivable qualities of these objects are their accidental qualities, these qualities may vary widely; therefore, *de facto* co-specific individuals may not appear to belong to the same species (think of biological species with morphologically very different sexes, for instance), while there may be very similar individuals apparently belonging to the same species, but in fact belonging to different ones (think of cases of near-perfect mimicry, etc.).

This last remark, however, immediately establishes the connection between the contemporary externalist Hilary Putnam's well-known "Twin Earth Experiment" and Ockham's conception: for just as on Putnam's account the rigidly designating terms 'H_2O' and 'XYZ' would have to have necessarily disjoint extensions even if the transparent liquid that is H_2O on this earth would be phenomenally indistinguishable from the similar liquid on twin-earth (whence the inhabitants of each planet would be willing to call both liquids 'water' in their miraculously matching English), so too, on Ockham's conception, rigidly designating natural kind terms designating distinct species would have to have disjoint extensions, still, on account of their phenomenal similarity, competent speakers of Latin may mistake members of one species for another.

Indeed, as Ockham and his ilk clearly realized in their speculations concerning *the possibility of perfect divine deception* on account of divine omnipotence, the occasional, naturally occurring and naturally detectable misjudgments might supernaturally be turned systematic, and in principle undetectable, which is precisely the core idea of the Cartesian Demon-scenario. And this is how Ockham's semantic externalism opens up the *apparent possibility* of an extreme version of "Demon-skepticism", which, however, as I have argued elsewhere and will argue soon here, is in fact *not* a genuine possibility. But to get to this point, let us first see exactly what this extreme version of "Demon-skepticism" involves.

In *The Matrix*, the celebrated movie premised on a brains-in-a-vat scenario, there is an interesting conversation among "the rebels", i.e., persons living in the devastated physical world of the twenty-second century, who originally acquired their concepts in the virtual reality of "the Matrix" (a computer program feeding artificially generated humans nurtured in complete sensory isolation from physical reality the virtual experiences of twenty-first century America as *we* know it). The conversation concerns what the artificial peptide goo served for dinner tastes like. The suggestions range from runny egg to Tasty Wheat to snot. But it soon turns out that the main concern *is not* that one of the interlocutors makes an *error in judgment* in the sense that what he deems, say, Tasty Wheat taste is really oatmeal, or chicken, or tuna taste. Rather, the concern is that the interlocutors literally have *no idea* of Tasty Wheat taste or chicken taste or tuna taste, or of genuine chickens or real tuna, for that matter. Having acquired their concept of, say, chickens in the virtual reality of the Matrix, in complete cognitive isolation from a real world that at least used to be populated by real chickens, this concept can represent only the virtual objects of this virtual reality, whatever those are, but *not* the real objects of physical reality.

In fact, Descartes had the same sort of concern when he worried *not only* about the possible *formal falsity*, as he called it, of our *judgments*, but also about the *material falsity* of our *simple ideas*, whether they are simple sensory or intellectual ideas, and whether these intellectual ideas are *adventitious*, empirical ideas, *acquired* somehow from sensory ideas, or *innate* intellectual ideas, co-created with our minds. For Descartes is not so much concerned about the sort of relatively easily corrigible error in judgment that stems from ordinary sensory illusion (after all, that's how we *know* about sensory illusion at all, namely, by our ability to detect it!), as about the systematic, *in principle incorrigible error* stemming from the *material falsity* of our simple ideas, whether they are *acquired* in a scenario of systematically deceptive quasi-experiences envisioned in the "dreaming argument" or *planted* in our minds by its maker or an omnipotent manipulator, as envisioned in the "demon argument".

Descartes' way out of the epistemic predicament of the Demon scenario is to show that it is not really possible: *at least one* of my simple ideas must be materially true, namely, the idea of myself; for otherwise I would have to accept the obviously self-defeating claims that I do not doubt, I do not think, I do not exist, whereas all these are refuted by the very act of doubting everything on account of thinking about the alleged possibility of the Demon scenario. Having thus proved, or rather, just realized, the material truth of the idea of *ego*, identified as the thinking self, Descartes can only claim real existence to this solipsistic "self", this lonely consciousness, which he, contrary to the prevailing Aristotelian Scholastic tradition, but apparently in line with the Augustinian-Platonic tradition, is quite happy to identify as his true self, the *res cogitans*, but which still cannot know anything about the real existence of anything else "outside" itself.

Indeed, this is precisely the key to understanding how it is the Demon-scenario that carves out from a full, flesh and blood Aristotelian-Scholastic human person, the anemic Cartesian self, thereby turning us into "ghosts in a machine"; this is why the Cartesian self is just the seat of consciousness, sitting in the middle of a theater of ideas it is immediately aware of, regardless of whether anything corresponds to them in external reality or not. But then, if we want to recover our genuine selves, we should first exorcise the Demon, but *not* through the solipsistic, introspective certainty of this lonely consciousness. In any case, this is the direction in which I am headed; the foregoing brief reminder of Cartesian Demon skepticism merely served to illustrate two important points regarding the very idea of Demon skepticism in general, in any of its actual varieties.

First, Demon skepticism concerns not only doubt concerning the formal truth of our judgments, but also doubt concerning the material truth of our simple ideas, or, in other words, the veridicality of our simple concepts, such as our concepts of natural kinds. Second, since according to Descartes, the formal truth of judgments entails the material truth of the ideas making them up; therefore, by contraposition, he takes it to be self-evident that the material falsity of our ideas entails the formal falsity of the judgments they make up, provided this material falsity is understood as the complete *failure of these ideas to engage reality* (to convey information about a mind-independent reality as it truly is in itself), which is precisely why I would call Descartes' materially false ideas *non-veridical* (not truth-telling) *concepts*.

In fact, semantically speaking, we may clarify the idea of a non-veridical concept's failure to engage reality by saying that in a formal semantics, categorematic terms expressing non-veridical concepts and those expressing veridical ones would take their semantic values from two disjoint sets, even if, perhaps, phenomenally, from the perspective of the minds that form these concepts, they may be indistinguishable. Thus, for instance, if I have the concept of chickens formed in physical reality upon encountering genuine chickens, then the concept I express by the word 'chicken' represents genuine chickens. On the other hand, if I was raised in the Matrix, what I can express by the word 'chicken' is at best a concept that represents virtual chickens, whatever those are, but definitely not chickens as we understand them. Still, the claim of the Demon argument is that I can have phenomenally the exact same mental contents whether I acquire my concepts in genuine or in virtual reality. So, the concept acquired in the Matrix would appear to me to represent the same in the same way as the concept acquired in genuine physical reality, despite the fact that only the latter represents chickens, and the former does not. It is *this* idea of a non-veridical concept that I would briefly describe as one that appears to represent something that it does not represent. But with *this* understanding of the idea of a non-veridical concept we can easily see that Ockham's externalism is committed to the possibility of the Demon-scenario, despite the fact that it is not a possibility.

3.4 Exorcizing the Demon *Without* an Appeal to Solipsistic Certainty

To see this in more detail, let us see why Ockham is in fact committed to this possibility. Given the way he describes how we acquire our concepts of natural kinds, Ockham has to admit that these concepts have their semantic contents independently from one another, and that their semantic content is merely contingently related to their phenomenal content, i.e., one can have the exact same concepts with the same phenomenal contents, even if they may, perhaps supernaturally, have wildly different semantic contents from what they actually have, which is precisely the gist of Ockham's externalism as Panaccio describes it: phenomenal content (which some might even call *intension*) does not determine semantic content (which again some may call *extension*). In fact, in Ockham's theory of concepts, the same observations would apply to all simple categorematic concepts. But then, this means that Ockham is committed to the claim that any of our simple categorematic concepts may possibly be non-veridical (i.e., they may be phenomenally the same whether they are acquired in actual physical reality or in the Matrix), and since given their simplicity they can be so *independently* of one another, it is at least logically possible that they are *all* non-veridical at the same time. But this is precisely how I would define a BIV, a brain-in-a-vat, to allude to Putnam's scenario, or I might as well call it the Cartesian solipsistic self, to allude to Descartes' scenario: a BIV is a cognitive subject that has no simple veridical concepts.[3]

As we can see, under this definition, Ockham's conception is committed to the claim that it is at least supernaturally possible that there is a cognitive subject, say, s, such that s is a BIV. Now, since s has phenomenally the same simple concepts as we do, and complex thoughts and judgments are semantically composed of these simple concepts, s and you and I are capable of thinking the same thoughts, in particular, the thought that s is a BIV.

Now let us assume that a mad scientist, or better yet, God, has made it to be the actual state of affairs that s is a BIV. So the judgment that s is a BIV is true. But an affirmative, contingent judgment, whoever forms it, can be true only if it is made up from veridical concepts (for otherwise it simply does not engage any real state of affairs, as I have argued earlier). Therefore, since by our assumption s is a BIV, s has no veridical concepts, and so his judgment that s is a BIV cannot be true. However, his judgment, being made up from the same concepts as ours, is the same as ours or anybody else's, which by our assumption was supposed to be true; therefore, we have to conclude that the judgment that s is a BIV is not true. Hence, together with the previous conclusion, the same judgment is both true and not true, which is an explicit contradiction.

However, since we have derived this contradiction from the assumption that s is a BIV with the help of other self-evident premises; the assumption cannot be true. But it is a consequence of Ockham's position that this assumption can be true; therefore, by *modus tollens*, Ockham's position cannot be true.[4]

In fact, by virtue of this reasoning, no position concerning concept-identity, whether it is internalist or externalist, which allows the possibility of the Demon-scenario as defined here can be true. Historically, I would say this is true of most of late-medieval and modern philosophy, the prominent representatives of which all worked under certain unquestioned assumptions, especially, under the assumption of the possibility of having the exact same concepts whether or not they actually manage to latch on to real, extramental objects, yielding the possibility of in principle undetectable, total deception about an external reality. But without even trying to justify this sweeping historical claim, I would rather contrast it with what I take to be an earlier paradigm, which simply does not allow the possibility of the Demon-scenario, and which is often referred to as Aquinas' doctrine of "the formal identity of knower and the known", but which, for the sake of better contrast, I would rather refer to as "the 'hyper-externalism' of Aquinas".

---

[3] Note that although I think the notion of a BIV *as defined here* is closely related to both Descartes' and Putnam's conceptions, it is not intended to capture *their* original intentions.

[4] For a formal statement and more detailed discussion of this argument, see my exchange with Claude Panaccio in Klima and Hall 2011.

3.5 The Hyper-Externalism of Aquinas and the Pervasiveness of Forms

As we have seen, Demon-skepticism in the sense defined here is possible only if our simple cognitive acts are merely contingently veridical, leaving open the possibility that perhaps *all* our simple, non-introspective cognitive acts are non-veridical. However, if a certain conception of the identity conditions of these cognitive acts demands that at least some of our simple non-introspective cognitive acts are essentially veridical, that is, their veridicality is part and parcel of their conditions of identity, then this conception directly excludes the possibility of Demon-skepticism without any appeal to the introspective certainty of a solipsistic Cartesian self.

The necessary veridicality of some of our simple cognitive acts with regard to their proper objects is a consequence of the Aristotelian idea that such a cognitive act is nothing but the form of the object in the cognitive subject in a different mode of existence. One way of demystifying this apparently rather obscure description is by appealing to the nowadays common idea of encoding and decoding, i.e., the process of transferring the same information through different media in a way that allows it to be reproducible in a numerically different copy. For instance, the recording and playback of a song provide an obvious case of this process. The song played back is a copy of the song originally played, where the reproduction of the song is possible by virtue of the preservation of the same information in the recording, which, in this sense, is but the form of the song originally played (the modulation of airwaves in the studio) in a different mode of existence, say, existing in the form of the pattern of tiny pits on the surface of a music CD encoding the modulation of airwaves.

Without arguing for it, let us just assume for the time being that this "demystification" correctly captures the original Aristotelian idea (originally illustrated by the example of a signet ring pressed into wax, when the wax takes on the form of the ring without its matter). However, even granting this perhaps dubious proposal, one may still have doubts whether it would yield the idea of the necessary veridicality of some simple cognitive acts with regard to their proper objects. After all, just as the pattern of pits on the surface of the CD could in principle be produced by something other than the recording apparatus without the original song actually played in the studio, so the same cognitive act could be produced in the subject without a "matching" object, rendering the act non-veridical, just as the Demon-scenario would suggest. So, apparently, the suggested "demystification" of the Aristotelian idea supports precisely the contingency of the veridicality of cognitive acts and thus the possibility of Demon-skepticism, contrary to what it was devised to illustrate.

However, to proceed from the better known to the lesser known, let us take a closer look at the case of the sound recording. The pattern of tiny pits on the surface of the CD is certainly producible by means other than the recording apparatus. After all, the same kind of laser beam with the same kind of modulation would produce the same pattern if the modulation of the laser beam were driven not by the modulation of electronic signals driven in turn by the modulation of airwaves hitting the microphones in the recording studio, but, say, by a computer producing the same modulation without any sound whatsoever. (Of course, this is the same sort of scenario as Putnam's ant producing what looks like a sketch of Churchill by crawling in the sand. Putnam 1981, p. 1).

However, and this is the important point, in that case the pattern of pits on the surface would *not* be a *record* of any sound whatsoever: it may be an ornament, it may be a surface feature, etc., but *not a record of some sound*. For the pattern of pits to qualify *as the record of a song*, it has to be part of the system of encoding and preserving information about the actual modulation of air vibrations constituting the song. Indeed, that for the *record of a song* as such it is *essential* to encode information about the song (whereas it is *accidental* that it is *this* pattern of pits in this system of encoding) is further confirmed by the fact that if I "rip" the track from the CD onto my computer's hard drive, then I get *the same song* onto my hard drive (for otherwise the RIAA would certainly have no business harassing me for pirating it), but now recorded in a different medium, this time encoded in the pattern of different magnetic polarities on the surface of the disk.

Describing this process in the language of Aristotelian hylomorphism, we can say that the form of the song that first informed the air in *esse reale*,[5] existing as the modulation of air waves, first was received in the matter of the CD in *esse intentionale*, without the matter it originally informed, merely coinciding with the pattern of pits informing the CD in *esse reale*; then again, it was received in the matter of the hard disk, in another instance of *esse intentionale*, again, without the matter of the original, this time coinciding with the pattern of polarities informing the disk in *esse reale*. Thus, in the whole process, what qualifies any real feature of any medium *as the record* or *encoding* of the original form is "the formal unity" of these real features in the sense that the system of encoding secures transferring and preserving *the same information* about the original form throughout the process. If the chain of transferring and preserving the same information is broken, and a merely accidentally similar pattern is produced by some other means, then it may be "misinterpreted" by the next decoder as a recording of some original, but it will never be *the same*, precisely because it does not fit into the chain in the same way, which is essential for the identity of any encoded bit of information. Thus, to switch to another example, even if a recorded TV program could not be distinguished from the live feed of the same by just looking at the screen, the two are *not the same*, and their difference *is* detectable precisely by looking at the process of the transfer of

[5] In this description, I am using the modern conception of sound, according to which it is nothing but the vibration of the air generated by a sounding object. So, on this conception, the term 'sound' refers to the modulation of the airwaves *in esse reale*. However, on the Aristotelian medieval conception, the term 'sound' would rather refer to a sensible quality of the sounding object that causes these airwaves, namely, its state of vibration that is encoded by the vibrations of the air. So, on this conception, the air vibrations would be just encodings (*species*) of the sound of the sounding object, whence the vibrations of the air would be nothing but the vibrations of the sounding object in *esse intentionale*. But this is just another way of saying that what *we* would call sound in the air is just an encoding of the sound of the sounding object, properly identifying the ultimate source of information in the process of encoding, transmitting, transcoding, and decoding information.

information producing the *exact same looking*, but *essentially different* images on the screen.

3.6 Rolling Back Our False Dilemmas

However, if on the strength of these examples we are willing to interpret the idea of formal unity between a cognizer and some cognized thing in the sense of the preservation of information, so that this is essential for the identity of the cognitive act insofar as it is a natural encoding of the form of the object, then it is not hard to see that those *simple* cognitive acts that are identified precisely in terms of receiving, storing and further processing information about their proper objects will have to be *essentially veridical*. For then these simple cognitive acts, regardless of what firing patterns of neurons in the brain or what spiritual qualities of an immaterial mind realize them, will count as the cognitive acts encoding information about their proper objects only if they do in fact represent those objects that they appear to represent to the cognitive subject, for they present or represent to the subject precisely the information they encode about their proper object.

Thus, on this conception, the veridical acts of perception, memory, and intellectual apprehension (as opposed to the non-veridical or contingently veridical acts of hallucination, imagination, dreaming, misremembering, judging, believing, etc.) are essentially, and not merely contingently veridical. But then, within this conception, the idea of "Demon skepticism" as described earlier is *ab ovo* excluded. Things *are* as they appear in our veridical acts of cognition, but sometimes, on account of the similarity of a veridical act of cognition to a non-veridical act or to a veridical act of cognizing something else, we may rashly judge things to be the way they appear to be through the non-veridical act, or to be that other thing. But since the veridical act is essentially veridical, and so it cannot be the same as a non-veridical act or the veridical cognition of something else, we can correct our mistake, by detecting the difference, as when we say, "Oh, I thought the bed was on fire, but it was just a dream" or "Oh, I thought I saw water on the road, but it was just a mirage".

But similar observations apply in the more elaborate cases. For instance, in the scenario of "the Matrix", the characters eating the peptide goo in physical reality have to realize that when they say it tastes like chicken, they have no genuine conception of chickens, as the only experiences they have about "chickens" are the virtual "chickens" of the Matrix. They could say they had a conception of chickens through those virtual experiences only if they could look at those virtual experiences as somehow carrying genuine information about genuine chickens, say, if whoever created the program had modeled the virtual chickens after real chickens and presented them as representations of real chickens in the way a nature video provides us with genuine information about genuine animals in remote lands. However if the virtual, quasi-experiences these people had in the Matrix are merely similar to genuine experiences, but *are not* genuine experiences (whether through direct perception or through "mediated perception", as through a documentary), then the concepts abstracted from those quasi-experiences *are not* the concepts of genuine things that would produce similar, but *never the same*, experiences. Thus, again, when it comes to the identity conditions of intellectual concepts, which on the Aristotelian account would carry just further processed, abstracted information about the genuine objects of genuine experiences, it is clear that on this conception they also have to be essentially veridical.

So, how is this conception related to the issue of externalism? As we could see, the way this conception identifies concepts has practically nothing to do with their internal or phenomenal properties: we talk about the same concept as long as it is a carrier of the same information whatever realizes it, and what determines this information is precisely the type of external object that the concept carries information about. Thus, from the perspective of this conception, whatever internal properties the concept has (say, whether it is a neural firing pattern of a certain type, etc.) is irrelevant, since the same concept, carrying the same information, can be realized in just any other type of "medium". Therefore, the internal properties of the concept not only do not fully determine its content, they have basically nothing to do with it; on this conception the content of the concept is fully externally determined, and so this conception can justifiably be dubbed "hyper-externalism". In any case, whatever we name it, it is clear that this conception does not allow the apparent possibility of the Demon-scenario.

But then, if it was this apparent possibility that allowed Descartes to "carve out" the mind as the conscious self, based on what he took to be the *only* necessarily veridical concept in the face of Demon skepticism, namely, the idea of this self itself, then abandoning this apparent possibility at once removes the artificial demarcation line between "the mental" and "the physical" stemming from this false assumption. We are biological organisms receiving, storing and further processing information about reality as it is, and not scared little ghosts looking at tiny pictures that appear to depict some reality, but who are reduced to mere guesswork about this reality without getting some divine guarantee concerning the reliability of at least some of these tiny pictures.

But still, saying that we are biological organisms need not land us on the materialist side of the fence, for it may well be the case, as Aquinas argued, that some phenomena of our intellectual lives, in particular, abstract concept-formation, still cannot take place in a material medium. So, what informs our material bodies may be informed by some forms of its own, in an activity of its own, which it may perform whether it is actually informing our bodies or not.

However, making better sense of this claim would get us embroiled in the *arcana* of Aquinas' metaphysics, denying nearly all our common, post-Cartesian intuitions.[6] All I hope to have shown here is that wanting to do so is not an entirely crazy idea, given our recalcitrant false dilemmas, based precisely on these usually unchecked intuitions.

[6] For some details of these *arcana*, see Klima 2007, 2009, 2015b.

References

Greco, J. 2004. Reid's reply to the Skeptic. In *The Cambridge companion to Reid*, ed. T. Cuneo and R. van Woudenberg. Cambridge: Cambridge University Press.

Klima, G. 2001. Existence and reference in medieval logic. In *New essays in free logic*, ed. A. Hieke and E. Morscher. Dordrecht/Boston/London: Kluwer Academic Publishers.

Klima, G. 2007. Thomistic 'Monism' vs. Cartesian 'Dualism'. *Logical Analysis and History of Philosophy* 10: 92–112.

Klima, G. 2009. Aquinas on the materiality of the human soul and the immateriality of the human intellect. *Philosophical Investigations* 32: 163–182.

Klima, G. 2011. Intentional transfer in Averroes, indifference of nature in Avicenna, and the representationalism of Aquinas. In *Universal representation, and the ontology of individuation, Proceedings of the society for medieval logic and metaphysics, 5*, ed. G. Klima and A. Hall. Cambridge: Cambridge Scholars Publishers.

Klima, G., and A. Hall. 2011. *The demonic temptations of medieval nominalism. Proceedings of the society for medieval logic and metaphysics, 9*. Newcastle upon Tyne: Cambridge Scholars Publishers.

Klima, G. 2015a. Semantic Content in Aquinas and Ockham. In *Linguistic content: New essays on the history of philosophy of language*, ed. M. Cameron and R.J. Stainton, 121–135. Oxford: Oxford University Press.

Klima, G. 2015b. Universality and immateriality. *Acta Philosophica* 24(2015): 31–42.

Panaccio, C. 2004. *Ockham on concepts*. Aldershot-Burlington: Ashgate.

Panaccio, C. 2015. Ockham's externalism. *Intentionality, cognition and mental representation in medieval philosophy*, ed. G. Klima, 166–185. New York: Fordham University Press.

Putnam, H. 1981. *Reason, truth, and history*. Cambridge: Cambridge University Press.

Chapter 4
Hylomorphism: Emergent Properties without Emergentism

William Jaworski

4.1 The Hylomorphic Notion of Structure

Hylomorphism claims that structure (or organization, form, arrangement, order, or configuration) is a basic ontological and explanatory principle. Some individuals, paradigmatically living things, consist of materials that are structured or organized in various ways. You and I are not mere quantities of physical materials; we are quantities of physical materials with a certain organization or structure. That structure is responsible for us being and persisting as humans, and it is responsible for us having the particular developmental, metabolic, reproductive, perceptive, and cognitive capacities we have.

The hylomorphic notion of structure is not the same as others that have appeared in the literature.[1] It is not the same, for instance, as the notion of structure that has been operative in discussions of grounding in metaphysics (Schaffer 2009; Sider 2012). 'Structure' in that sense refers to what grounds the distinction between things that are fundamental or perfectly natural in Lewis' (1983) sense, and things that are not. Hylomorphism provides one account of what structure in this sense includes, but the specifically hylomorphic notion of structure is not the same as this one. Nor is the hylomorphic notion of structure the same as the notion that is operative in debates about scientific realism (Worrall 1989; Ladyman and Ross 2007). 'Structure' in those debates refers to the relational contents of scientific theories that remain constant across episodes of theory change, but that is not what 'structure' in the

[1]Other accounts of hylomorphic structure that have appeared in the literature include Fine's (1999), Johnston's (2006), Oderberg's (2007), Koslicki's (2008), and Rea's (2011). The account of hylomorphic structured I'll be developing here differs from theirs in ways that I trust will become evident as I proceed.

W. Jaworski (✉)
Fordham University, Bronx, NY, USA
e-mail: jaworski@fordham.edu

M. García-Valdecasas et al. (eds.), *Biology and Subjectivity*,
Historical-Analytical Studies on Nature, Mind and Action 2,
DOI 10.1007/978-3-319-30502-8_4

hylomorphic sense refers to. Nor is the hylomorphic notion of structure the same as the one David Chalmers sometimes employs (2002, p. 258). For Chalmers, structural descriptions are abstract microphysical descriptions of a system's state at a time that are contrasted with microphysical dynamic descriptions of how a system's states change over time. For hylomorphists, by contrast, structure is not an abstract postulate, nor are descriptions of structures confined to microphysics.

The hylomorphic notion of structure is closer to the notion of organization that biologists often appeal to. Here is one example taken from a popular college-level biology textbook—note the references to organization, order, arrangement, and related things:

> Life is highly organized into a hierarchy of structural levels… Biological order exists at all levels… [A]toms… are ordered into complex biological molecules… the molecules of life are arranged into minute structures called organelles, which are in turn the components of cells. Cells are [in turn] subunits of organisms… The organism we recognize as an animal or plant is not a random collection of individual cells, but a multicellular cooperative… Identifying biological organization at its many levels is fundamental to the study of life… With each step upward in the hierarchy of biological order, novel properties emerge that were not present at the simpler levels of organization… A molecule such as a protein has attributes not exhibited by any of its component atoms, and a cell is certainly much more than a bag of molecules. If the intricate organization of the human brain is disrupted by a head injury, that organ will cease to function properly… And an organism is a living whole greater than the sum of its parts… [W]e cannot fully explain a higher level of order by breaking it down into its parts (Campbell 1996, pp. 2–4).

This passage suggests that the way things are structured, organized, or arranged plays an important role in them being the kinds of things they are, and in explaining the kinds of things they can do.

To help illustrate the hylomorphic notion of structure I'll use a simple example; we can call it *the squashing example*. Suppose we put Godehard in a strong bag—a very strong bag since we want to ensure that nothing leaks out when we squash him with several tons of force. Before the squashing the contents of the bag include one human being; after they include none. In addition, before the squashing the contents of the bag can think, feel, and act, but after the squashing they can't. What explains these differences in the contents of the bag pre-squashing and post-squashing? The physical materials (whether particles or stuffs) remain the same—none of them leaked out. Intuitively we want to say that what changed was the way those materials were structured or organized. That organization or structure was responsible for there being a human before the squashing, and for that human having the capacities it had. Once that structure was destroyed, there no longer was a human with those capacities. Structure is thus a basic ontological principle: it concerns what things there are. It is also a basic explanatory principle: it concerns what things can do.

Another feature of structure is brought out by the neurophysiologist Miller:

> [T]he physical universe tends towards a state of uniform disorder… In such a world the survival of form depends on… [either] the intrinsic stability of the materials from which the object is made, or the energetic replenishment and reorganisation of the material which is constantly flowing through it… The configuration of a fountain… is intrinsically unstable, and it can retain its shape only by endlessly renewing the material which constitutes it; that

> is, by organising and imposing structure on the unremitting flow of its own substance… The persistence of a living organism is an achievement of the same order as that of a fountain… it can maintain its configuration only by flowing through a system which is capable of reorganising and renewing the configuration from one moment to the next. But the engine which keeps a fountain aloft exists independently of the watery form for which it is responsible, whereas the engine which supports and maintains the form of a living organism is an inherent part of its characteristic structure (1978, pp. 140–141).

Many structures, including those that distinguish living things from nonliving ones, are not static spatial relations, such as the relatively unchanging spatial relations among atoms in a crystal; they are instead dynamic patterns of environmental interaction—what Johnston (2006) calls 'dynamic principles of unity' (Fine 1999, pp. 68–69, calls them 'principles of variable embodiment'). They comprise programmatic sequences of changes over time, and often involve different kinds of changes under different kinds of conditions. It is because of their dynamic structures, their ability to impose those structures on incoming matter and energy, that complex individuals (paradigmatically living things) persist through the constant influx and efflux of matter and energy that characterize their interactions with the wider world.

The foregoing remarks gesture toward a view of structure in the natural world. We can express that view by describing the theoretical roles structure is supposed to play:

> **Structure matters:** it operates as an irreducible ontological principle, one that accounts at least in part for what things essentially are.
>
> **Structure makes a difference:** it operates as an irreducible explanatory principle, one that accounts at least in part for what things can do, the powers they have.
>
> **Structure counts:** it explains the unity of composite things, including the persistence of one and the same living individual through the dynamic influx and efflux of matter and energy that characterize many of its interactions with the wider world.

A final idea about structure is introduced by Dewey. He suggests that mental phenomena can be understood as species of structural phenomena in general:

> The difference between the animate plant and the inanimate iron molecule is not that the former has something in addition to the physico-chemical energy; it lies in the *way* in which physico-chemical energies are interconnected and operate… Iron as a genuine constituent of an organized body acts so as to tend to maintain the type of activity of the organism to which it belongs. If we identify… the physical as such with the inanimate we need another word to denote the activity of organisms… Psycho-physical is an appropriate term… Psycho-physical does not denote an abrogation of the physico-chemical; nor a peculiar mixture of something physical and something psychical… it denotes the possession of certain qualities and efficacies not displayed by the inanimate. Thus conceived there is no problem of the relation of physical *and* psychic. There are specifiable empirical events marked by distinctive qualities and efficacies. There is first of all, *organization*… Each 'part' of an organism is itself organized, and so of the 'parts' of the part… '[M]ind' is an added property assumed by a feeling creature, when it reaches that organized interaction with other living creatures which is language, communication (1958, pp. 253–58).

If structure is uncontroversially part of the natural world, and mental phenomena are just species of structural phenomena, then they must be uncontroversially part of the natural world as well. If Dewey is right, then structure plays a fourth theoretical role:

Structure minds: it provides us with resources for understanding the place of mental phenomena within the natural world.

Knowing this much about the theoretical roles hylomorphic structure is supposed to play it's possible to get a rough preliminary sense for how hylomorphists will approach mind-body problems. Hylomorphic structure carves out distinctive individuals from the otherwise undifferentiated sea of matter and energy that is or will be described by our best physics, and it confers on those individuals distinctive powers. If hylomorphic structure exists, then the physical universe is punctuated with pockets of organized change and stability—composite physical objects (paradigmatically living things) whose structures confer on them powers that distinguish what they can do from what unstructured materials can do. Those powers include the powers to think, feel, perceive, and act.

A worldview that rejects hylomorphic structure, by contrast, lacks a basic principle which distinguishes the parts of the physical universe that can think, feel, perceive, and act from those that can't, and without a basic principle that carves out zones with distinctive powers, the existence of those powers in the natural world can start to look inexplicable and mysterious. If there is nothing built into the basic fabric of the universe that explains why Zone A has powers that Zone B lacks—if nothing explains why you, say, have the power to think, feel, and perceive, while the materials surrounding you do not, then the options for understanding the existence of those powers in the natural world become constrained: either those powers must be identified with the powers of physical materials taken by themselves or in combination (as panpsychists and many physicalists claim), or their existence must be taken as an inexplicable matter of fact (as many emergentists and epiphenomenalists claim), or their existence in the natural world must be denied altogether (as substance dualists and eliminative physicalists claim). If there is hylomorphic structure, however, the options are no longer constrained in this way.

On the hylomorphic view, distinctive powers like yours and mine exist in the natural world because structure exists in the natural world. Moreover, because structure is a basic principle on the hylomorphic view, this does not simply push the demand for an explanation back a step. A framework's basic principles stand in need of no further explanation. Structures and things that get structured are both basic on the hylomorphic view. Nothing must explain why the former exist any more than something must explain why the latter do. As a result, the view leaves it unmysterious why and how thought, feeling, perception, and intentional action exist in the natural world.

To make the hylomorphic view more than a mere suggestion, we need to describe it in detail, and to do that, it will be helpful to have a working metaphysic.

4.2 Sparse Properties and Powers

Let's assume a substance-attribute ontology that takes substances or *individuals*, as I'll call them, and attributes or *properties*, to be fundamental entities. The properties I have in mind are *natural* properties, not mathematical or logical ones. For our purposes we can put properties of the latter sorts to one side. Natural properties are causal-enablers: they confer the powers that make causal interactions among individuals possible. They are also causal-explainers: they explain why individuals act or are acted on in the ways they are. Natural properties also ground the objective similarities and differences among individuals. Individuals are always similar or dissimilar in certain respects, and these respects are properties (Martin 1996a, pp. 71–73, 1997, 2007, pp. 42–43).

The view I've described takes properties to be *sparse* not *abundant* (Lewis 1983). They do not correspond to predicates one-one. Predicates can be identified with sentence-frames, linguistic expressions such as '___ is red' and '___ is taller than ___' that form sentences when the blanks are filled in by terms (Strawson 1974, pp. 37–38; Armstrong 1978a, pp. 2–3). Properties are supposed to be the non-linguistic correlates of at least some predicates. If properties are sparse, then it is possible for different predicates to express the same property ('weighs 453.59 grams', 'weighs 1 pound'), and for different properties to be expressed by the same predicate in different contexts ('The team is good', 'The wine is good') (Armstrong 1978b, pp. 9–14; Campbell 1990, p. 25; Molnar 2003, p. 26). Likewise, not every predicate will express a property. For one thing, there might be unknown properties to which no actual predicates correspond (Armstrong 1978b, pp. 12–14; Molnar 2003, p. 25). In addition, some predicates are self-referentially incoherent; they do not (and indeed cannot) correspond to any sparse property (2003, p. 26). It is also possible to invent predicates, but it is not possible to invent sparse properties. For all of these reasons and others, sparse properties do not correspond one-one to predicates.

What, then, determines which predicates express sparse properties? What moreover determines which properties those predicates express, and whether they are the same or different properties that are expressed by other predicates? According to the account of properties I've been outlining, the answers to these questions derive largely from empirical sources. Roughly, we take our best empirical descriptions, explanations, methods, and techniques, and countenance all the properties needed to make the descriptions and explanations true and the methods and techniques effective (Swoyer 1982, p. 205; Ellis 2002, pp. 44–45; Molnar 2003, p. 27; Armstrong 2010, p. 19).[2]

Properties are also powers. There are many competing theories of powers in the literature. The one I favor is a version of the theory defended by Martin (1996a, b, 1997, 2007), Heil (2003, 2005), and Martin and Heil (1998, 1999). I'll follow Heil,

[2] Brian Ellis states the basic idea as follows: "[W]hich predicates designate properties[?]… I answer: first decide what properties and structures you must postulate if you wish to give an adequate account of the phenomena, and then decide which expressions of the language refer to these properties or structures" (2002, pp. 44–45). Consider likewise Armstrong, Molnar, and Swoyer respectively: "[*H*]ow do we determine what the true universals are? My suggestion is that they are best postulated on the basis of *total science* (2010, p. 19); "[W]hat properties there are is… determined… on a posteriori grounds, most likely by current best science" (2003, p. 27). "the claim that there are such things as properties is a philosophical one, but determining just what properties there are is – like questions about existence generally – an empirical matter" (1982, p. 205).

and call it the identity theory of powers (Jaworski 2014, 2016). According to the identity theory of powers, properties are essentially dispositional; each essentially empowers its individual possessor to interact with other individuals in various kinds of ways. A diamond's hardness empowers it to do a variety of things—to scratch glass, for instance. We describe this power-conferring role in a many different ways. We say that the diamond is hard, that the diamond is able (or has the power or potential or capacity) to scratch glass, or that the diamond would scratch that mirror if raked across its surface. These different vocabularies create the impression that there are different kinds of properties: dispositional and categorical (or qualitative). According to the identity theory, however, these vocabularies describe the very same properties; they just bring out the different theoretical roles these properties play. Dispositional descriptions such as 'The diamond would scratch that mirror if raked across its surface' bring out the roles it plays as a power. Nondispositional descriptions such as 'The diamond has a tetrahedral arrangement of carbon atoms' bring out its role as a stable manifestation, or actualization, of the power the carbon atoms have to be arranged tetrahedrally. The one property is thus simultaneously both a stable manifestation of a power and a power itself, both an actuality and a potentiality.

The identity theory of powers has several noteworthy features. First, it claims that powers are essentially directed toward their manifestations. This directedness has led some philosophers to draw analogies between dispositionality and intentionality (Martin and Pfeifer 1986; Place 1996a, b; Molnar 2003). Intentional mental states are said to be directed at things. My desire is essentially a desire *for* something, my fear is essentially a fear *of* something. Something analogous is true of powers; they are essentially powers *for* various manifestations. The property of fragility, for instance, is essentially directed toward breaking.

Likewise, just as my desire can remain unfulfilled and my fear unrealized, so too a power can remain unmanifested. A quantity of table salt has the power to dissolve in water, but it might never actually be dissolved, and a fragile vase might never actually break.[3]

Another feature of the identity theory is that powers are manifested only in specific circumstances and typically only in conjunction with individuals that have reciprocal powers—what Martin calls 'reciprocal disposition partners'. Powers can be manifested both actively and passively: both in the ways individuals affect things and in ways they are affected by them. Powers are manifested only when individuals with reciprocal powers are conjoined in the right circumstances. Water, for instance,

[3] Martin (1996a) defends this idea with an example: it seems possible that there might be fundamental physical particles in the universe that have the power to interact in various ways with particles around here, and yet that are so far away that they reside outside the light cone of the particles around here. The two groups of particles never actually interact, yet it seems obvious that the distant particles still have the power to interact with the local ones.

can exercise its power to dissolve things only in conjunction with things that have the power to be dissolved by it.[4]

In addition, the same power can manifest itself differently in conjunction with different disposition partners. To use Heil's example: a ball will roll on a hard surface on account of its roundness, and it will make a concave depression in a soft surface on account of that same roundness. The same property, the ball's roundness, manifests itself in different ways in conjunction with different disposition partners. Likewise, the diamond's hardness empowers it to scratch glass and also to scratch jade, and the batter's power to hit a baseball 400 feet also empowers him to hit a bigger, heavier softball 300 feet.

4.3 Individual-Making Structures

The ontology I've just described provides a basis for understanding the hylomorphic notion of structure. Structures are supposed to confer powers. The squashing example, for instance, suggests that structure is what confers the powers to think, feel, perceive, and act. In the ontology I've just outlined, it is properties that confer powers. So in this ontology, structures are properties. If that's the case, then structures have all the characteristics of properties.

First, structures are powers—powers in particular to configure (or organize, order, or arrange) things. Each structured individual organizes or configures the materials that compose it. I configure the materials that compose me, and you configure the materials that compose you. Describing the way each of us configures our respective materials is something that hylomorphists say is an empirical undertaking—in our cases, an undertaking left largely to biology, biochemistry, neuroscience, and other biological subdisciplines.

Second, structures have the same directedness that all powers do. The structures of living things in particular appear to be directed toward developing and maintaining organisms' mature states, as well as the powers that characterize those states and their manifestations (Mayr 1997, p. 22). Like many of the powers we have considered, an organism's structure manifests itself in many different ways—in, for instance, the organism's various developmental stages, as well as in the variety of self-regulating processes that maintain the cells, tissues, and organs that living things develop such as photosynthesis, glycolysis, protein synthesis, and so on.

[4] Harré and Madden's (1975) examples of radioactive decay and ammonium tri-iodide seem initially to provide counterexamples to the general rule that powers are manifested or exercised in pairs, or triples, or *n*-tuples. But even here it might be possible to understand the cases in a way that conforms to the general reciprocity model. At the very least the environment surrounding the radioactive nuclei or the ammonium tri-iodide cannot include any agents that inhibit the exercise of their powers to decay or explode, respectively. Environments that are free of inhibitory factors might then be viewed as reciprocal disposition partners for the decaying nuclei and the explosive compound.

With these points in mind consider again the theoretical roles attributed to structures earlier. It should be apparent how structure makes a difference on the hylomorphic view. Structures are powers. Consequently, if something has a structure, it has powers that it would not otherwise have. Structure thus makes a difference. Structure also matters. Structured individuals have their configuring properties essentially; each is essentially an organizer/configurer of materials. For a structured individual to cease configuring some materials or other is for that individual to cease to exist. Structures are thus essential properties of structured individuals. In addition, unlike many of the powers we have considered, structures cannot go unmanifested. Crystals of table salt can sit idly; their power to dissolve in water can remain forever unmanifested. But there is no sitting idle when it comes to my power to configure the materials that compose me. If I am not manifesting that power, if I am not configuring those materials, then I do not exist, and if I do not exist, there is no individual to do my configuring. Structures, then, are not just essential powers of structured individuals; they are powers of structured individuals that are essentially manifested, that cannot exist unmanifested.

Structure also counts; it explains the unity and persistence of composite things, and in the case of living things that means explaining their unity and persistence through the dynamic influx and efflux of matter and energy that characterizes their interactions with the surrounding world. To understand how structure plays this role it is helpful to consider the hylomorphic view of composition.

4.4 Hylomorphic Composition

The hylomorphic view of composition is similar in its outlines to Peter van Inwagen's (1990). Van Inwagen presents his view as an answer to the Special Composition Question (SCQ): Under what conditions do multiple things compose one thing? According to van Inwagen, composition happens exactly if the activities of some fundamental physical particles constitute a life. By 'a life' van Inwagen takes himself to mean what biologists do: "the individual life of a concrete biological organism… [the] sense according to which 'Russell's life' denotes a purely biological event" (1990, p. 83). Van Inwagen's descriptions of lives stay largely at the level of metaphor and analogy. The reason is that providing the literal details of what lives are and what characteristics they have is, he thinks, an empirical undertaking (1990, p. 84). It is the job of biology to tell us what life is, he says, just it is the job of chemistry to tell us what an acid is.

Van Inwagen's view implies that *x* is a proper part of *y* if and only if *y* is an organism and *x* is caught up in the life of *y* (1990, p. 94). The expression "caught up in a life" is one that van Inwagen borrows from the biologist Young (1971). Van Inwagen explains with an example:

> Alice drinks a cup of tea in which a lump of sugar has been dissolved. A certain carbon atom… is carried along with the rest of the sugar by Alice's digestive system to the intestine. It passes through the intestinal wall and into the bloodstream, whence it is carried to

> the biceps muscle of Alice's left arm. There it is oxidized in several indirect stages (yielding in the process energy… for muscular contraction) and is finally carried by Alice's circulatory system to her lungs and there breathed out as a part of a carbon dioxide molecule… Here we have a case in which a thing, the carbon atom, was… caught up in the life of an organism, Alice. It is… a case in which a thing became however briefly, a *part* of a larger thing when it was a part of nothing before or after… (1990, pp. 94–95).

According to van Inwagen, then, composition does not happen apart from lives; composite beings are all living things.

Let the foregoing remarks suffice for an overview of van Inwagen's account of composition. Importantly, van Inwagen's lives play precisely the kinds of theoretical roles that hylomorphic structures are supposed to play. *Lives matter* on van Inwagen's view; they are ontological principles: whether the *x*s constitute a life makes a difference to whether or not some composite individual exists. Likewise, *lives make a difference*; they are explanatory principles: living things are capable of doing things that cannot be exhaustively described and explained using the conceptual resources used to describe and explain the materials that compose them (1990, pp. 122, 180). Finally, *lives count*; they operate as principles of unity (p. 121) and persistence (pp. 145, 148): what binds the simples that compose me into a single being is that their activity constitutes a life, and what enables me to persist through changes in those simples is the persistence of that life. Because van Inwagen's lives play these roles it is easy to use his view of composition as a basis for developing the hylomorphic view.

Configuring materials and being composed of materials are co-foundational concepts on the hylomorphic view, just as having a life and being composed of simples are co-foundational concepts on van Inwagen's. Likewise, just as van Inwagen restricts composition to living things, hylomorphists restrict it to structured individuals in general. According to hylomorphists, composition occurs when and only when an individual configures materials; there is a *y* such that the *x*s compose *y* if and only if *y* is an individual that configures the *x*s. I will call individuals that configure the materials composing them *structured individuals*.

Structured individuals are *emergent* individuals on the hylomorphic view. There are empirically-describable conditions that are sufficient to bring into existence new structured individuals where previously no such individuals existed. How do we know on this account which arrangements of physical materials correspond to actual structures and which are mere arrangements? Here again hylomorphists take a cue from van Inwagen. Structured individuals have non-redundant causal powers that mere arrangements of physical materials do not have. The chair, the mountain, and the planet don't do anything that cannot be exhaustively described and explained by appeal to physical materials alone, but living things are capable of doing things that can only be done by unified composite individuals. We are thus forced to grant that there are such individuals (1990, pp. 118, 122).

Once a structured individual comes into existence it is essentially and continuously engaged in configuring materials. The materials it configures are precisely those that compose it. When it comes to characterizing the configuring activity of structured individuals, hylomorphists can adopt most of what van Inwagen says about lives, at least when it comes to the configuring activities of living things, the paradigmatic structured individuals. My life is identical to my configuring various fundamental physical materials at various times—an event that has the characteristics van Inwagen attributes to lives, and that has many other characteristics it is business of the biological sciences to describe. An individual living thing does not configure exactly the same materials for very long since those materials are in constant flux, yet despite this, the individual maintains itself one and the same through all the changes on account of its ongoing configuring activity. That activity is what unifies various materials into a single individual, both synchronically and diachronically, just as lives do on van Inwagen's account.

The hylomorphic view has many of the same implications as van Inwagen's. Like van Inwagen's view, for instance, it is committed to property pluralism. It implies that structured individuals have properties of at least two sorts: properties due to their structures (or their integration into individuals with structures), and properties due to their materials alone independent of the ways they are structured. Subatomic particles, atoms, and molecules have physical properties such as mass irrespective of their surroundings. Under the right conditions, however, they can contribute to the activities of living things. Nucleic acids, hormones, and neural transmitters are examples. They are genes, growth factors, and metabolic and behavioral regulators. Each admits of two types of descriptions. They can be described in terms of the contributions they make to a structured system, but they are also independently describable in non-contribution-oriented terms. Descriptions of the former sort express the properties characteristic of structured individuals such as organisms and their parts. Descriptions of the latter sort express the properties they possess independent of their integration into structured wholes. A strand of DNA might always have various atomic or fundamental physical properties regardless of its environment, but it acquires new properties when it is integrated into a cell and begins making contributions to the cell's activities. It becomes a gene, a part of the cell that plays a role in, say, protein synthesis.

Some philosophers and biologists call the new properties acquired by structured systems *emergent properties*. Emergent properties have three characteristics:

1. They are first-order properties, not higher-order ones; that is, they are not logical constructions with definitions that quantify over other properties.
2. They are not epiphenomenal, but make distinctive causal or explanatory contributions to the behavior of the individuals having them.
3. They are possessed by an individual on account of its organization or structure.

Notice: it is not a characteristic of emergent properties (at least not on the hylomorphic view) that they are generated or produced by lower-level systems. I'll return to this point later on.

There are also some noteworthy differences between the hylomorphic view and van Inwagen's. In the interests of time, I'll mention only one. The hylomorphic view of parts is less revisionary than van Inwagen's. Van Inwagen (1981, 1990) denies that there are organic parts such as eyes, hearts, and kidneys; the only proper parts are fundamental physical particles and individual cells. Hylomorphists, by contrast, accept that there are parts such as eyes, hearts, and kidneys. Their reasons for doing so are broadly empirical. If we accept a broad naturalism in matters of ontology, then empirical adequacy is an important criterion for determining which principles of part identity and individuation to adopt. And there is some reason to think that our best descriptions and explanations of human behavior postulate parts like these.

Even though it is possible to divide a human along, say, spatial lines into thirds, or fifths, or tiny metric cubes, biologists, neuroscientists, and psychologists are typically more interested in dividing humans and other organisms along functional lines (Bechtel 2007, 2008; Craver 2007).[5] Perhaps the best example of how the biological sciences divide organisms into hierarchically-ordered functional parts is provided by the method of *functional analysis*.[6]

Functional analysis is a method that biologists, cognitive scientists, engineers, and others frequently employ to understand how complex systems operate. It involves analyzing the activities of those systems into simpler subactivities performed by simpler subsystems (Fodor 1968; Cummins 1975; Lycan 1987, Chapter 4; Bechtel and Richardson 1993; Glennan 2002; Bechtel 2007, 2008; Craver 2007).[7] Consider a complex human activity such as running. Functional analysis reveals that running involves a circulatory subsystem that is responsible for supplying oxygenated blood to the muscles. Analysis of that subsystem reveals that it has a component responsible for pumping the blood—a heart. Analysis of the heart's pumping activity shows that it is composed of muscle tissues that undergo frequent contraction and relaxation. These activities can be analyzed into the subactivities of various cells, and these in turn can be analyzed into the operations of various organelles. This analytic process continues until something is discovered to have a property or to engage in an activity not on account of the activities of some lower-level subsystems but as an unanalyzable matter of fact. At that point, functional analysis comes to an end.

Functional analysis is important because it provides a basis for understanding the kinds of parts postulated by descriptions and explanations in the biological sciences. Those parts are subsystems or components, things that contribute in empirically-specifiable ways to the activities of the wholes to which they belong. Saying that *x* is a part of *y* implies that *y* engages in some activity that is composed of subactivities of the sort that can be revealed through functional analysis, and that *x* performs one

[5] Craver (2007, Chapter 5) calls purely spatial parts 'pieces' and parts in the functional sense 'components'. Heil (2003, p. 100) also suggests something like the distinction between merely spatial parts and parts of other sorts, which he calls 'substantial parts'.

[6] The term 'functional analysis' is due to Cummins (1975). Becthel (2008) calls it 'mechanistic decomposition' or 'functional decomposition'. Craver (2007) subsumes it under the heading of 'mechanistic explanation'. He takes Cummins's notion of functional analysis to be the exemplar of what he calls the 'systems tradition', but argues that Cummins fails to provide an adequate account of mechanistic explanation. He thus distances himself from the term 'functional analysis'.

[7] Elsewhere I've discussed functional analysis in greater detail. I've argued among other things that it does not correspond to the notion of function that is operative in discussions of functionalism in the philosophy of mind (including teleological functionalism), and that it does not imply a commitment to reductionism (Jaworski 2011, 2012, 2016).

of those subactivities. Saying that my heart is part of me is saying that my activities are composed of subactivities (and those perhaps of further sub-subactivities, and so on), and that one of those subactivities (or sub-subactivities or sub-sub-…subactivities) is performed by the heart. What distinguish different parts of me from each other, moreover, are the different ways they contribute to my activities: different parts contribute in different ways.

Because hylomorphism endorses this account of parthood, its inventory of parts is less revisionary than van Inwagen's. It is able to countenance the functional parts postulated by the biological sciences, and in many cases these will include parts that are postulated by common sense as well.

4.5 Activity-Making Structures and Embodiment

My discussion of structure has so far focused on individual-making structures, the structures that make individuals what they are. But on the hylomorphic view I been describing, these aren't the only structures that exist. The activities in which structured individuals engage have structures as well—activity-making structures. These structures provide us with resources for understanding how structure minds.

The idea that there are activity-making structures is based on the observation that the activities of structured individuals involve coordinated manifestations of their powers and the powers of their parts. The activities of structured wholes are not random assortments of physiological occurrences. Walking, talking, throwing, driving, swimming, eating, breathing, and the other activities we engage in involve some of our parts manifesting their powers in organized or coordinated ways. We engage in these activities precisely by imposing an order on the ways our parts manifest their powers. Doing so does not bring into existence new individuals but rather activities.

The coordinated manifestation of powers we find in activities is a species of structuring, another manifestation of the power organisms have of imposing order on materials. My parts needn't manifest their powers in an ordered way. It is possible for my neurons to fire or my muscles to contract in ways that do not compose an activity of throwing a baseball, or swimming, or playing an instrument. Fatigue, injury, insufficient training, and many other factors can result in uncoordinated manifestations of the powers of my parts. But when I succeed in throwing, swimming, or playing, I succeed in imposing a structure on the way my parts manifest their powers; I structure their manifestations throwing-, swimming-, or playing-wise.

A large number of activities in which we engage also involve the coordinated manifestation of the powers of surrounding individuals and materials. Throwing a baseball, swimming, and playing an instrument, for instance, all have environmental requirements. They require a baseball, water, and an instrument, respectively. Those things, moreover, must be properly disposed; the baseball must be free to travel when I release it, for instance; it cannot be taped or glued to my hand. Living things like us are thus zones of structured activity that extend beyond the materials that compose us.

Activity-making structures have characteristics similar to those of the individual-making structures discussed earlier. Activity-making structures matter. Structured activities have their activity-making structures essentially. When I throw a baseball my movements are essentially coordinated in the way it takes for me to engage in that activity. If my parts and the surrounding materials had some other activity-making structure, if they were configured playing-an-instrument-wise, they would compose an entirely different activity. Activity-making structures are thus ontological principles; they matter; they account in part for what the activities they configure essentially are.

Activity-making structures also make a difference. They are powers. If an individual engages in a structured activity, that activity confers on the individual powers that it would not otherwise have. If I am engaged in the activity of throwing a baseball—if my parts and the surrounding materials are structured throwing-a-baseball-wise—then I have the power to make an out in a game or to knock down cans at the county fair, and these are powers I would not possess if my behavior were not structured this way. Activity-making structures thus make a difference.

Activity-making structures also count; they confer unity on diverse events in something analogous to the way individual-making structures confer unity on the physical materials that compose individuals. My parts and the surrounding materials needn't manifest their powers throwing-a-baseball-wise. The very same muscle fibers that contract in my shoulder when I throw a baseball might also contract when I experience an uncontrolled muscle spasm, or when a physical therapist stimulates them electrically. What unifies the contractions of the muscle fibers, what coordinates the manifestations of their powers is what I do when I try to make an out, or try to knock down cans at the county fair, or try to accomplish whatever I try to do when I throw a baseball. In undertaking these activities I impose a unified order on the way my parts and the surrounding materials manifest their powers.

Earlier I described the unifying role that individual-making structures play in terms of composition. It is possible to articulate an analogous notion of composition for activities. Just as physical materials compose an individual exactly if the right kind of individual-making structure is imposed on them, likewise various events compose an activity exactly if the right kind of activity-making structure is imposed on them.

Here is one attempt to define a notion of activity composition: Suppose that a is a structured individual with the power to engage in activity F. Suppose, moreover, that $b_1, b_2, \ldots, b_n$ are individuals, and that a subset of the bs are proper parts of a. Suppose, finally, that each b_i has the power to engage in an activity, G_i: b_1 has the power to G_1, b_2 has the power to G_2, and so on. We can then define a notion of activity composition as follows:

Activity composition: Necessarily, a's F-ing at t is composed of b_1's G_1-ing at t_1, b_2's G_2-ing at $t_2, \ldots$ and b_n's G_n-ing at t_n exactly if b_1's G_1-ing at t_1, b_2's G_2-ing at $t_2, \ldots$ and b_n's G_n-ing at t_n are structured F-wise, where time t includes the times $t_1, t_2, \ldots, t_n$.

Although this formulation looks rather messy, the idea is very simple: *a* engages in the activity of *F*-ing exactly if *a* imposes an *F*-wise organization on the ways its parts and surrounding materials manifest their powers. I throw a baseball exactly if I coordinate the ways my parts and surrounding materials manifest their powers throwing-a-baseball-wise.

Some notes about this definition: First, given reasonable assumptions, activity composition implies that the behaviors of structured individuals never violate the laws governing their fundamental physical constituents. According to hylomorphism, the activities of structured wholes are composed of the structured manifestations of the powers of their lower-level constituents and surrounding materials. If those constituents or materials were to lose their powers, or were to become incapable of manifesting them, they would become incapable of composing the activities of structured wholes. Those activities depend on lower-level items retaining and manifesting the powers they have. By analogy, it is only because bricks and timbers retain their shapes under compression that they can be recruited as components of buildings. Similarly, it is only because lower-level materials retain their distinctive powers that structured individuals can recruit them as components for their own activities. This is one thing that sets the hylomorphic view apart from many emergentist theories which claim that higher-level powers trump or nullify the powers of lower-level things.

Second, this definition of activity composition implies a way of understanding the idea that a structured individual has its powers *in virtue of* having its parts. A structured individual has the power to engage in various activities because it has parts. Those parts form a subset of the individuals with powers whose coordinated manifestations compose its activities. We can also express this idea by saying that the powers of a structured individual are *embodied in* its parts. My visual system embodies my power to see; your circulatory system embodies your power to bring oxygenated blood to various parts of yourself; Gabriel's nervous system embodies his power to coordinate the movements of his limbs, and so on.

According to the hylomorphic theory I've been describing all the powers of structured individuals are essentially embodied in their parts; the activities in which it engages are essentially composed of the coordinated manifestations of the powers of its parts and surrounding materials. It is impossible, not just nomologically, but metaphysically, for me to engage in the activity of throwing a baseball unless my parts and the surrounding materials manifest their powers in a coordinated way. On the hylomorphic view I've been describing, the same is true of all our activities. We can call this the *embodiment thesis*.[8]

---

[8] Many hylomorphists of the past have denied the embodiment thesis. Aristotle himself appears to deny it in *De Anima* III.4. There he appears to argue that understanding or *nous*, the power to grasp the essences of things, has no organ and is in general unmixed (*amigēs*) with a body (429a 10–27). There are several things to say in response. First, a commitment to the essential embodiment of our capacities is the default position for a hylomorphist. In fact Aristotle treats embodiment as the default position as well (403a 16–19, 24–27; 403b 17–18). We cannot do most of the things we do (walking, breathing, perceiving, eating, and so on) without some of our parts manifesting their powers in coordinated ways. Against this background of embodied activity the claim that *nous* is

If the embodiment thesis is true, then thinking, feeling, perceiving, intentionally acting, and other phenomena that philosophers typically categorize as mental are all coordinated manifestations of the powers of our parts and surrounding individuals and materials. When I think, feel, perceive, or act, I coordinate the manifestation of the powers of my parts and in some cases the powers of surrounding individuals and materials as well. Those coordinated manifestations compose my thoughts, feelings, perceptions, and actions. An example will help illustrate this idea.

Perceptual states are often categorized as mental states. According to hylomorphists, when Gabriel sees something—a table, say, he and the table both manifest powers they possess. He manifests the power to see the table, and it manifests the power to be seen by him. Gabriel and the table are reciprocal disposition partners. The powers of both are mutually manifested in each other's presence when the surrounding conditions are right, just as water and salt mutually manifest their powers to dissolve and be dissolved when conditions are right.

Moreover, Gabriel has the power to see by virtue of having the parts he has. The coordinated manifestations of the powers of some of his parts (intuitively those composing his visual system) contribute to his seeing, and those parts form a subset of the individuals whose powers, when manifested in the right way, compose his seeing. The same is true mutatis mutandis of the table. Intuitively, the parts of the table by virtue of which it has the power to be seen are those composing its surface, the ones which reflect light to Gabriel's eyes. In addition, there are other environmental factors involved in Gabriel's seeing the table such as the direction and intensity of the light, the condition of the air through which he sees it, and so on. Gabriel's seeing the table is thus a complex structured activity composed of the coordinated manifestation of the powers of his parts and those of the surrounding materials. Details about what parts those are, what powers they have, and how the manifestations of those powers must be structured in order to compose our activities are all to be supplied empirically through functional analysis.

4.6 Naturalistic, Antireductive, and Unmysterious

The picture of mental phenomena I've sketched is naturalistic, antireductive, and unmysterious. Moreover, the view is not physicalist in at least this sense: it denies that everything can in principle be exhaustively described and explained by physics. In particular, the structures that make living beings or mental beings what they are cannot be described exhaustively using only the conceptual resources of physics.

not essentially embodied stands out as anomalous. Commentators like Shields suggest that the claim stretches Aristotle's hylomorphic framework "almost beyond the limits of recognition" (2014, pp. 354–355). Given the foregoing, a hylomorphist should look to reject the embodiment thesis only if there is a powerful argument against it. Most of the arguments that have been advanced against the thesis are descendants of the argument in *De Anima* III.4. I argue elsewhere in detail that that argument is flawed in a variety of ways (Jaworski 2016).

Nor is this limitation merely a function of our peculiar descriptive and explanatory interests, as nonreductive physicalists are wont to claim, for on the hylomorphic view, structure is something that is fundamentally different from things that get structured. Let me say a word about each of these features.

The view is naturalistic, I would suggest, because it claims that we are physical beings with physical components, and that our distinctive powers to think, feel, perceive, and act are essentially embodied in the physical materials that compose us. It also implies robust necessitation and supervenience theses. Here are four that are entailed by the embodiment thesis; although in the interests of time I won't reproduce the proofs here (for details see Jaworski 2016: chapter 9):

> **Hylomorphic activity necessitation:** Necessarily, for any structured individual, x, if x engages in activity A at time t, then **(a)** x has proper parts and surrounding materials with powers such that the A-wise manifestation of those powers at t composes x's A-ing at t, and **(b)** necessarily, for any individual z and time t^*, if z has proper parts and surrounding materials at t^* that are exactly similar to x's at t, then z engages in A-ing at t^*.
>
> **Hylomorphic power necessitation:** Necessarily, for any structured individual, x, if x has the power to engage in activity A at time t, then **(a)** x has proper parts with powers such that the A-wise manifestation of those powers in conjunction with the powers of surrounding materials at a time, t^*, would compose x's A-ing at t^*, and **(b)** necessarily, for any individual z and time t^*, if z has proper parts at t^* exactly similar to x's at t, then z has the power to engage in A-ing at t^*.
>
> **Hylomorphic activity supervenience:** For any possible worlds w_1 and w_2, and any individual x in w_1 and individual y in w_2, if the activities of x at time t and y at time t^* are composed of the structured manifestations of the powers of exactly similar parts and surrounding materials, then y will engage in activities at t^* exactly similar to the activities of x at t.
>
> **Hylomorphic power supervenience:** For any possible worlds w_1 and w_2, and any individual x in w_1 and individual y in w_2, if x at time t and y at time t^* have exactly similar parts and surrounding materials, then y will have powers at t^* exactly similar to the powers of x at t.

Roughly, these theses imply that if I engage in an activity A at time t, or have the power to engage in A-ing at t, then any individual that has parts exactly like mine at t^*, and whose surrounding environment is exactly like mine at time t^* will engage in A or have the power to engage in A-ing at t^*. There can be no differences between my powers and activities and those of another individual without there being some difference in our parts or surrounding materials.

Two further necessitation and supervenience theses are not entailed by the embodiment thesis, but they follow from it given reasonable assumptions:

> **Structo-physical necessitation:** Necessarily, for any x, if x is a structured individual at time t, then **(a)** there are physical materials, $y_1, y_2,\ldots, y_n$, that compose x at t, and **(b)** necessarily, for any physical materials, $z_1, z_2,\ldots, z_n$, and time t^*, if the zs at t^* are exactly similar to the ys at t in respect of those properties and relations that can be exhaustively described by physics, then they compose an individual at t^* that is structurally exactly similar to x at t.
>
> **Structo-physical supervenience:** For any possible worlds w_1 and w_2, and any physical materials, $x_1, x_2,\ldots, x_n$, in w_1 and $y_1, y_2,\ldots, y_n$ in w_2, if the xs at time t are exactly similar to the ys at time t^* in respect of the kinds of properties and relations that can be exhaustively

described by physics, then the *x*s compose an individual, *z*, at *t* if and only if the *y*s compose an individual at *t** that is structurally exactly similar to *z* at *t*.

Roughly, these theses say that physical twins must be structural twins. Any structural difference between me and another individual must be correlated with a physical difference between us. There can be no structure zombies—no cases in which physical materials are interacting in exactly the ways they are in me without those materials composing an individual that is structured exactly similar to me.

Describing why the view is antireductive is a somewhat lengthy endeavor. This isn't the feature I want to focus on, so I'll just briefly describe the reasoning behind it. Roughly, reduction consists in descriptive and explanatory takeover (Jaworski 2011, 2012, 2016). To say that a theoretical framework T1 is reducible to another framework T2 is to say that T2 can take over all the descriptive and explanatory jobs that T1 performs. We've seen that the hylomorphic view implies a commitment to property pluralism. The properties things have on account of their structures are different from the ones their constituent materials have on their own, independent of any structure. Because of this, structural descriptions and explanations are irreducible to descriptions and explanations that appeal to unstructured materials for reasons that William Bechtel summarizes (though his preferred term is 'organization'):

> [T]he organization of… components typically integrates them into an entity that has an identity of its own… Organization itself is not something inherent in the parts… In virtue of being organized systems, mechanisms do things beyond what their components do… As a result, organized mechanisms become the focus of relatively autonomous disciplines… This autonomy maintains that psychology and other special sciences study phenomena that are outside the scope of more basic sciences but which determine the conditions under which lower-level components interact. In contrast, the lower-level inquiries focus on how the components of mechanisms operate when in those conditions… The fact that mechanisms perform different activities than do their parts manifests itself in the fact that the activities of whole mechanisms are typically described in different vocabulary [sic] than are component operations… The vocabulary used in each science describes different types of entities and different operations – one describes the parts and what they do, whereas another describes the whole system and what it does (2007, pp. 174, 185–186).

Someone might worry that the view I've described can't be antireductive. Hylomorphism implies a commitment to necessitation and strong supervenience, after all, and, someone might urge, these theses pose a threat to the autonomy of higher-level explanations. If all of my activities are necessitated by the properties of my parts and surrounding materials, then, says the worry, there is little or no explanatory work to be done by any properties I possess as a composite whole such as my thoughts and feelings.

But hylomorphists argue that the premise is false. Neither necessitation nor supervenience threaten the efficacy or autonomy of higher-level explanations. The idea that they do is typically based on a misguided assumption about the nature of necessitation and supervenience relations, the assumption that necessitation and/or supervenience entail some type of explanatory condition. According to this assumption, saying that *A*-properties are necessitated by or supervene upon *B*-properties implies that *B*-properties explain *A*-properties. But necessitation and supervenience merely state patterns of property covariation; they do not purport to explain why the patterns obtain, nor do they imply that higher-level conditions obtain because of lower-level ones (Kim 1984, 1990; Grimes 1988; Jaworski 2016).

Determination relations, by contrast, conjoin necessitation with explanation. Lower-level determination in particular claims that the activities of a whole are both necessitated and explained by the activities of its parts and surrounding materials:

> **Lower-level determination:** Necessarily, for any structured individual, *x*, if *x* engages in activity *A* at time *t*, then (a) *x* has proper parts and surrounding materials **with powers, *P1*, *P2*,..., *Pm*, such that the manifestations of *P1*, *P2*,..., *Pm* at *t* explain *x* 's *A* -ing at *t***, and (b) necessarily, for any individual *z* and time *t**, if *z* has proper parts and surrounding materials at *t** exactly similar to *x*'s at *t*, **with powers, *Q1*, *Q2*,..., *Qm*, exactly similar to the powers of *x* 's proper parts and surrounding materials**, then *z* engages in *A*-ing at *t**, **and the manifestations of *Q1*, *Q2*,..., *Qm* at *t** explain *z* 's *A* -ing at *t****.

This definition is the same as the definition of hylomorphic activity necessitation except for the addition of the boldface clauses, which express explanatory relations. Lower-level conditions not only necessitate higher-level ones, they explain them as well. If lower-level determination is true, all the activities of a structured whole are explained by the activities of its parts and surrounding materials. In that case, lower-level conditions might pose a threat to the efficacy or autonomy of higher-level conditions for reasons that are well-rehearsed in the literature (Kim 1998).

But hylomorphists reject lower-level determination. It might be plausible to endorse something like lower-level determination when it comes to the structure-independent properties of a whole. The mass of a structured whole is plausibly determined by the masses of its fundamental physical constituents since those constituents have a collective mass irrespective of how they are arranged. But hylomorphists deny that all the properties of structured wholes are like this. Some properties are structure-dependent or emergent; they depend not just on the materials composing a thing, but on the way those materials are structured or organized. The higher-level activities of a structured whole might covary in all metaphysically possible worlds with certain lower-level material conditions, but those lower-level conditions do not explain why the structured whole engages in the higher-level activities it does, at least not in the way exponents of lower-level determination envision. The reason is that the manifestations of the powers of a structured individual's parts and surrounding materials at a time do not by themselves compose the activities in which a structured individual engages. The manifestations of the powers of those parts and surrounding materials must be structured in the right way as well.

Finally, I've claimed that the hylomorphic view leaves it unmysterious how mental phenomena can exist in the natural world. I've explained how mental phenomena are species of structural phenomena. The hylomorphic view takes thoughts, feelings, perceptions, and the like to be structured activities, and it understands structured activities in terms of a general account of structure. So if structure is an unmysterious citizen of the natural world, then mental phenomena should be as well. How might we make the case for this?

One way, perhaps, is to appeal to *ontological naturalism*, the idea that when it comes to determining what exists, empirical investigation—paradigmatically science—is our best guide. We can think of ontological naturalism as the conjunction of a broadly Quinean thesis about ontological commitment with a broad empiricism. The broadly Quinean thesis maintains that we are committed to all the entities postulated by our best descriptions and explanations of reality, and a broad empiricism maintains that our best descriptions and explanations of reality derive from empirical sources such as the natural and social sciences. I've indicated that many scientists make use of a notion of structure or organization very similar to the hylomorphic one. If we are committed to the existence of the entities needed to make our best empirical descriptions and explanations true, then we have good empirical reasons for committing ourselves to structure—and hence, given the other things I've said, good reasons to find the existence of mind in the natural world unmysterious.

Of course, someone weaned on the problem of emergence might demand an explanation of how on the hylomorphic view lower-level physical occurrences manage to generate or produce higher-level mental phenomena such as consciousness. But such a demand misunderstands the hylomorphic view (Jaworski 2011, pp. 352–353, 2016). A request to explain why or how it is the case that *p* assumes that it is in fact the case that *p*. The request that hylomorphists explain why or how lower-level physical occurrences generate thoughts, feelings, and perceptions assumes that lower-level physical occurrences do in fact generate thoughts, feelings, and perceptions. But on the hylomorphic view, that assumption is false. Structured things are not in general produced by the lower-level things they structure. A piano, for instance, is not a produced by the wood and metal that compose it; it is produced, rather, by someone who imposes a piano-like arrangement on those materials. Similarly, on the hylomorphic view, thoughts, feelings, and perceptions are structured activities. They are not produced by the lower-level physical states that compose them; they are produced, rather, by someone who imposes a thinking, feeling, or perceiving structure on those states—who coordinates the way his or her parts manifest their powers. Because lower-level states do not in fact generate higher-level mental states on the hylomorphic view, it is illegitimate to request an explanation for why or how this happens, just as it would be illegitimate to request that a meteorologist explain why or how the will of Zeus produces rain.

I've argued that hylomorphism provides a nonphysicalist view of thought, feeling, perception, and action that is naturalistic, antireductive, and unmysterious. If these are attractive features in a mind-body theory, then hylomorphism is an alternative to physicalism that is worth taking seriously. There is much more that needs to be said about the hylomorphic view I've described, but I hope I've said enough to encourage you keep structure in mind.[9]

[9] A version of this paper was presented at the Fordham-Rutgers Metaphysics of Mind Conference held at Fordham University in March 2015. I discuss the hylomorphic view it outlines in greater detail in *Structure and the Metaphysics of Mind: How Hylomorphism Solves the Mind-Body Problem* (Oxford University Press, 2016). Some of the materials that appears in this chapter was originally published in that book, and has been reproduced by permission of Oxford University Press https://global.oup.com/academic/product/structure-and-the-metaphysics-of-mind-9780198749561.

References

Armstrong, D.M. 1978a. *Realism and nominalism: Universals and scientific realism*, vol. 1. Cambridge: Cambridge University Press.
Armstrong, D.M. 1978b. *A theory of universals: Universals and scientific realism*, vol. 2. Cambridge: Cambridge University Press.
Armstrong, D.M. 2010. *Sketch for a systematic metaphysics*. Oxford: Oxford University Press.
Bechtel, W., and R.C. Richardson. 1993. *Discovering complexity: Decomposition and localization as strategies in scientific research*. Princeton: Princeton University Press.
Bechtel, W. 2007. Reducing psychology while maintaining its autonomy via mechanistic explanations. In *The matter of the mind*, ed. M. Schouten and H. Looren de Jong. Oxford: Blackwell Publishing.
Bechtel, W. 2008. *Mental mechanisms: Philosophical perspectives on cognitive neuroscience*. London: Routledge.
Campbell, K. 1990. *Abstract particulars*. Oxford: Basil Blackwell.
Campbell, N.A. 1996. *Biology*, 4th ed. San Francisco: The Benjamin/Cummings Publishing Company, Inc.
Chalmers, D.J. 2002. Consciousness and its place in nature. In *Philosophy of mind*, ed. D.J. Chalmers. Oxford: Oxford University Press.
Craver, C.F. 2007. *Explaining the brain: Mechanisms and the mosaic unity of neuroscience*. Oxford: Oxford University Press.
Cummins, R. 1975. Functional analysis. *Journal of Philosophy* 72: 741–764.
Dewey, J. 1958. *Experience and nature*. New York: Dover Publications.
Ellis, B. 2002. *The philosophy of nature: A guide to the new essentialism*. Chesham: Acumen.
Fine, K. 1999. Things and their parts. *Midwest Studies in Philosophy* 23: 61–74.
Fodor, J. 1968. The appeal to tacit knowledge in psychological explanation. *Journal of Philosophy* 65: 627–640.
Glennan, S. 2002. Rethinking mechanistic explanation. *Philosophy of Science* 69: S342–S353.
Grimes, Th.R. 1988. The myth of supervenience. *Pacific Philosophical Quarterly* 69: 152–160.
Harré, R., and E.H. Madden. 1975. *Causal powers: A theory of natural necessity*. Oxford: Basil Blackwell.
Heil, J. 2003. *From an ontological point of view*. Oxford: Clarendon.
Heil, J. 2005. Dispositions. *Synthese* 144: 343–356.
Jaworski, W. 2011. *Philosophy of mind: A comprehensive introduction*. Malden: Wiley-Blackwell.
Jaworski, W. 2012. Structures, powers, and minds. In *Powers and capacities in philosophy: The new Aristotelianism*, ed. R. Groff and J. Greco. London: Routledge.
Jaworski, W. 2014. Hylomorphism and the metaphysics of structure. *Res Philosophica* 91: 179–201.
Jaworski, W. 2016. *Structure and the metaphysics of mind: How hylomorphism solves the mind-body problem*. Oxford: Oxford University Press.
Johnston, M. 2006. Hylomorphism. *Journal of Philosophy* 103: 652–698.
Kim, J. 1984. Concepts of supervenience. *Philosophy and Phenomenological Research* 45: 153–176. Reprinted in Kim 1993, 53–78.
Kim, J. 1990. Supervenience as a philosophical concept. *Metaphilosophy* 21: 1–27. Reprinted in Kim 1993, 131–60.
Kim, J. 1993. *Supervenience and mind Cambridge*. New York: Cambridge University Press.
Kim, J. 1998. *Mind in a physical world*. Cambridge, MA: MIT Press/Bradford Books.
Koslicki, K. 2008. *The structure of objects*. Oxford: Oxford University Press.
Ladyman, J., and D. Ross. 2007. *Every thing must go: Metaphysics naturalized*. Oxford: Oxford University Press.
Lewis, D.K. 1983. New work for a theory of universals. *Australasian Journal of Philosophy* 61: 343–377.
Lycan, W.G. 1987. *Consciousness*. Cambridge, MA: MIT Press.

Martin, C.B. 1996a. Properties and dispositions. In *Dispositions: A debate*, ed. T. Crane. London: Routledge.
Martin, C.B. 1996b. Replies to Armstrong and Place. In *Dispositions: A debate*, ed. T. Crane. London: Routledge.
Martin, C.B. 1997. On the need for properties: The road to Pythagoreanism and back. *Synthese* 112: 193–231.
Martin, C.B. 2007. *The mind in nature*. Oxford: Oxford University Press.
Martin, C.B., and J. Heil. 1998. Rules and powers. *Philosophical Perspectives* 12: 238–312.
Martin, C.B., and J. Heil. 1999. The ontological turn. *Midwest Studies in Philosophy* 23: 34–60.
Martin, C.B., and K. Pfeifer. 1986. Intentionality and the non-psychological. *Philosophy and Phenomenological Research* 46: 531–554.
Mayr, E. 1997. *This is biology: The science of the living world*. Cambridge, MA: The Belknap Press of Harvard University.
Miller, J. 1978. *The body in question*. New York: Random House.
Molnar, G. 2003. In *Powers: A study in metaphysics*, ed. S. Mumford. Oxford: Oxford University Press.
Oderberg, D.S. 2007. *Real essentialism*. New York: Routledge.
Place, U.T. 1996a. Intentionality as the mark of the dispositional. *Dialectica* 50: 91–120.
Place, U.T. 1996b. Dispositions as intentional states. In *Dispositions: A debate*, ed. T. Crane. London: Routledge.
Rea, M.C. 2011. Hylomorphism reconditioned. *Philosophical Perspectives* 25: 341–358.
Schaffer, J. 2009. On what grounds what. In *Metametaphysics: New essays on the foundations of ontology*, ed. D. Chalmers, D. Manley, and R. Wasserman. Oxford: Oxford University Press.
Sider, T. 2012. *Writing the book of the world*. Oxford: Oxford University Press.
Strawson, P.F. 1974. *Subject and predicate in logic and grammar*. London: Methuen.
Swoyer, Ch. 1982. The nature of natural laws. *Australasian Journal of Philosophy* 60: 203–223.
Van Inwagen, P. 1981. The doctrine of arbitrary undetached parts. *Pacific Philosophical Quarterly* 62: 123–127.
Van Inwagen, P. 1990. *Material beings*. Ithaca: Cornell University Press.
Worrall, John. 1989. Structural realism: The best of both worlds? *Dialectica* 43: 99–124.
Young, J.Z. 1971. *An introduction to the study of man*. Oxford: The Clarendon Press.

Chapter 5
Remarks on the Ontology of Living Beings and the Causality of Their Behavior

Thomas Buchheim

5.1 Two Arguments for the Ontological Difference Between Corporeal and Psychic Reality

Aristotle developed two good arguments for why one cannot identify psychic or mental states with bodily states or somatic processes (for instance neural activity in the brain) that are still broadly used today. The first argument states that things that are the subject or substrate of psychic states are essentially different to those that can be the subject of bodily states or somatic processes. The second argument says that between a psychic phenomenon and a simultaneous bodily process no isomorphic structural outline and therefore no analytically productive mapping can be undertaken.

To begin with the first argument: while *all* bodies taken simply as such, no matter how large or small, are subjects of somatic processes and bodily states (as already implied by the term "body"), only a very small selection of them can function as subjects of psychic states, and in these later cases, only when each is *taken as a*

This chapter was originally written for the conference 'Mind-brain: biology and subjectivity in contemporary Philosophy and Neurosciences' (48th Reuniones Filosóficas 2011) at the University of Navarra in Pamplona. The outlines of the proposed concept of mental causality were firstly published (in German) in *Philosophisches Jahrbuch* 119 (2012), 330–346 and subsequently discussed and developed in volume 120 (2013), 101–173 and 371–393 of the same journal. A former version in English was published in: Miroslaw Szatkowski & Marek Rosiak (Ed.): *Substantiality and Causality*, Berlin, New York: De Gruyter, 2014 pp. 21–34. The presently published version, besides some corrections and enlargements by the author, has been benefitted a great deal by the careful revision of Nathaniel Barrett. I wish to thank him for this piece of labor. I also wish to thank Marcela García for her translation of the original German paper into English and for her helpful and competent discussion of its ideas.

T. Buchheim (✉)
Department of Philosophy, Ludwig-Maximilians-University of Munich, Munich, Germany
e-mail: thomas.buchheim@lrz.uni-muenchen.de

M. García-Valdecasas et al. (eds.), *Biology and Subjectivity*, Historical-Analytical Studies on Nature, Mind and Action 2,
DOI 10.1007/978-3-319-30502-8_5

whole, and only if they fulfill extremely restrictive conditions regarding their internal structure and external demarcation. Aristotle very deliberately refuses to assume (as Plato and Descartes did) that psychic phenomena have an extra *psychic* subject of their own – for instance, a "res cogitans" or "soul" as a separately existing entity, because he clearly recognizes that such states (everything that we call "psychic" or "mental") only appear *as embedded in the vital context* of living individuals who, as far we can see, are themselves complex bodies.

Although these complex bodies, according to Aristotle, have a soul or are animate, the soul is not an independent entity besides the complex body, but rather a special manner of existing that this body has as a whole:

> [**1**] Hence the rightness of the view that the soul cannot be without a body, while it cannot *be* a body. That is why it is *in* a body, and a body of a definite kind. It was a mistake, therefore, to do as former thinkers did, merely to fit it into a body without adding a definite specification of the kind or character of that body [...] It comes about as reason requires: the fulfillment (*entelecheia*) of any given thing can only be realized in what is already capable of being that thing, i.e. in a matter of its own appropriate to it. From all this it is plain that soul is a fulfillment or account (*logos*) of something that possesses a capability of being such (*de An.* II 2, 414 a 19–28. Translations of Aristotelian texts are drawn from Aristotle 1984).

Psychic and mental states can only be ascribed to this totality in its way of existing: the *whole* human (and only the whole) thinks; the *whole* deer (only the whole) scents something; the *whole* blackbird (only the whole) sings. All mental and psychic states are biographical in nature. That is, they only appear as *embedded in life episodes* of living individuals in their totality. As long as these individuals are corporeal, the ultimate subject of psychic and mental states is indeed—although complex and highly integrated—a *body*. This is at least Aristotle's position.

But the bodies that can be, on the one hand, subjects of somatic or physical states and processes and, on the other hand, subjects of psychic or mental states are essentially *different*. While *all* bodies and bodily parts in any partition serve for the first kind (and the segregation of 'one' of them from the other and its surroundings is a matter of arbitrariness), only those highly complex bodies in so far as they are alive can be considered for the second kind, their segregation from the environment being definite by its living unity.

For this reason, in one and the same complex body there can be states of both kinds: first, physical states, and these can apply thoroughly in all parts of the body; and second, psychic states, but only in the body as a *whole* and therefore always embedded in life episodes of the whole individual. In support of this view of Aristotle that is, in my opinion, also an exemplary position for contemporary debates, I want to mention briefly two quotes from *De anima*, although there are many other passages in the philosopher's work that are equally relevant:

> [**2**] Yet to say that it is the soul which is angry is as if we were to say that it is the soul that weaves or builds houses. It is doubtless better to avoid saying that the soul pities or learns or thinks, and rather to say that it is the man who does this with his soul (*de An.* I 4, 408 b 11–18).

> [**3**] The view we have just been examining, in company with most theories about the soul, involves the following absurdity: they all join the soul to a body, or place it in a body, without adding any specification of the reason of their union, or of the bodily conditions required for it. Yet such explanation can scarcely be omitted; for some community of nature is presupposed by the fact that the one acts and the other is acted upon, the one moves and the other is moved; but it is not the case that *any* two things are related to one another in these ways. All, however, that these thinkers do is to describe the specific characteristics of the soul; they do not try to determine anything about the body which is to contain it, as if it were possible, as in the Pythagorean myths, that any soul could be clothed in any body [...] (*de An.* I 3, 407b15-24).

We can call this first argument for the difference between psychic and bodily processes in short the argument of different selectivity of the subjects: while psychic states select exactly one body with a certain inner complexity and definite demarcation against its environment, bodily processes in principle always diffuse into their surroundings and are selected under the artificial aspects of an observer.

In the course of contemporary philosophical debate with neuroscience a similar criterion was again formulated by Bennett and Hacker in their book *The Philosophical Foundations of Neuroscience* (2003). The authors mentioned referred to the false conclusion that stems from ignoring this argument as the "mereological fallacy," (Hacker, Bennett 2003, pp. 68ff.) which consists in the inconsequential transition from the activities of the whole organism to those processes related to them in some relevant part (such as brain regions) of this organism. According to this fallacy, it seems as though our brain (or certain parts of it) 'thinks,' 'sees,' or 'feels.' We begin to ask ourselves how it is possible that besides electrochemical neuro-activity there are other states, namely psychic or mental ones, in the brain or elsewhere, or whether we should assume that a 'person,' 'soul,' or 'mind' is something completely different in us that accomplishes these activities (seeing, thinking, and feeling).

The second argument, which was also brought up by Aristotle, refers to the interior structure of life episodes and the psychic states and processes that are embedded in them, which seems to be completely different to the structure of bodies.

While in the case of bodily processes, the individual, momentary states of the parts involved actually make up the whole process accumulated spatiotemporally, in the case of life episodes we cannot dissect the whole event into spatiotemporal details and moments without losing something. Rather, the episode or psychic state (for instance an itch or a sight impression) disappears completely when we go below a certain extension of the spatiotemporal regions in which it appears (cf. e.g. Stump 1999, pp. 417 f). This can be expressed as follows: life episodes appear in space and time but cannot be sharply distinguished spatiotemporally in the same way as their bodily symptoms. However, if I'm not mistaken, this kind of distinguishability belongs essentially to all *corporeal* states and processes. Aristotle concluded that life episodes or psychic events could not themselves be presented or understood as somatic complexes, but require a peculiar scientific approach and treatment. I quote a related passage from the work *On Generation and Corruption*:

> [**4**] An additional absurdity is that the soul should consist of the elements, or that it should be one of them. How are the soul's alterations to take place? How, e.g., is the change of being musical to being unmusical, or how is memory or forgetting, to occur? For clearly, if

> the soul be Fire, only such properties will belong to it as characterize Fire *qua* Fire; while if it be compounded, only the corporeal modifications will occur to it. But the changes we have mentioned are none of them corporeal. The discussion of these difficulties, however, is a task appropriate to a different investigation (*GC* II 6, 334a9-15).

What Aristotle says here about the 'matter of the soul' or fire, as was supposed at the time, also holds good mutatis mutandis today for the very much discussed 'soul matter' of neural activity in our brain. This activity, in its somatic profile, cannot account for the properties and sequences of activity that we discover in psychic phenomena within the vital context, like for instance meaning something (intentionality) or to feel a certain way (experience quality). It's difficult to see how to interpret what actually seems to happen, as embedded in a life episode (like remembering something), as a somatic process. According to Aristotle, we should rather do what is still common practice today, namely, speak of a mere *correlation* between what happens somatically and what happens in a life episode[1]; a correlation in which the spatiotemporal interior differentiation of both sides clearly cannot be mapped onto each other. This is why one side behaves as a somatic complex but the other one doesn't.

> [**5**] the alteration of that which undergoes alteration is also caused by the above-mentioned characteristics, which are affections of some underlying quality. Thus we say that a thing is altered by becoming hot or sweet or thick or white; and we make these assertions alike of what is inanimate and of what is animate, and further, where animate things are in question, we make them both of the parts that have no power of sense perception and of the senses themselves. For in a way even the senses undergo alteration, since actual perception is a motion through the body (*dia tou sômatos*) in the course of which the sense is affected in a certain way (*Phys*. VII 2, 244 b 6–12).

Aristotle seems to be very conscious of the fact that the change of states in a series of life episodes is not spatiotemporally distinguishable in the way a sequence of bodily states is. He determines therefore that the first only appears 'through' or with support of the second, although it is not to be identified with it. Compare a famous passage in *De Anima*:

> What we mean is not that the movement is *in* the soul, but that sometimes it terminates in the soul and sometimes starts from it, sensation e.g. coming from without, and reminiscence starting from the soul and terminating with the movements or states of rest in the sense organs (*De An*. I 4, 408b15-18).

---

[1] The correlation of psychic states and brain states is itself a product of theories and cannot be an immediate object of experiments; immediately determinable are the correlations between short term behavior events and cerebral activity. On this topic see Kurthen 2006, especially p. 28: "This is the structure of these experiments: primarily we have the investigation of a relation between an instruction and a behavior; secondarily, on the one hand, relations between the instruction and those thoughts it involves against the background of cognitive science and, on the other hand, a relation between a behavior event and those brain processes that accompany this event (or even just their indicators, as in an fMRI experiment). The relation between thoughts (the actual phenomena of sensation in the monkey, the actual episode of memory in the human) and brain processes appears at least in a tertiary perspective – and therefore at a highly theoretical level, not directly at the level of 'data' – as the relation between the supposed mental 'component' of a behavior and the brain processes measured directly or indirectly."

In other words, all the spatiotemporal differences which make up movement or alteration of bodies according to Aristotle do not really apply to the characteristic features which mark correlated psychic states. So bodily processes and psychic phenomena are not mapping to each other in a distinctly traceable way.

Indeed, that which definitely has different properties (here: different spatiotemporal distinguishability) cannot be identical according to Leibniz's principle of identity. Nevertheless, it doesn't follow for Aristotle that psychic or life episodes belong to a non-corporeal substance of its own. Rather, the substance to which they belong is a certain complex body that simultaneously exemplifies different kinds of states.[2] Every psychic event (thought, sensation, feeling) is, according to Aristotle, a culmination (climax, emphasis) of being alive that the whole system generates. That is, it is a particular character within the framework of certain life-episodes, which in turn are *versions* of its being alive as a whole. Even mere life—for instance, in sleep or unconsciousness—requires the coordinated activity of the whole body and its parts: metabolism, circulation, muscular tone, nervous dispositions, etc.

A sensation, a pain, a dream, a calculation, catching a ball or keeping balance are climaxes or *culminations* of being alive sustained by this active basis. Just as the mere being-alive relates to the integrated detailed states of the whole body and its parts, so does the culmination relate to the integrated *variations* of those detailed states of the whole system. This is what Aristotle calls self-accomplishment or fulfillment (*energeia* or *entelecheia*). Every accomplishment of a psychic function is such a culmination or an operative state of the whole system. But both, *prôtê* and *eschatê entelecheia* (cf. *Metaph.* IX 8, 1050a24) are embedded in life episodes and require therefore the whole complex body in its interior organization, the whole living individual.

5.2 The Somatic as Symptom of the Psychic

Psychic states are thus necessarily understood as states in their own right; but they do not belong as such to an incorporeal substance of their own. They exist in correlation with certain bodily states of the *same* corporeal complex substance as a

[2] This was reformulated by Peter F. Strawson as an argument for the special ontological status of self-conscious persons, but is valid in a reduced form for all living systems from a certain level of development: "What I mean by the concept of a person is the concept of a type of entity such that *both* predicates ascribing states of consciousness *and* predicates ascribing corporeal characteristics, a physical situation etc. are equally applicable to a single individual of that single type" (1959, p. 101 f.). Strawson has certain reasons to apply the concept of person primitively and exclusively to individuals with self-consciousness; he especially rejects the analysis of person as "animated body" or "embodied anima," as if these were simpler and the person were composed of these. At the same time it seems necessary to use the double aspect of such predicates also for non-personal living beings without having to consider the "animated body" as a composition of body and soul.

whole.[3] This relation I call horizontal dualism, because it is *not* a vertical relation of *layering* two different substances and their different classes of properties, but rather an intertwining of two orders of the same material in one single substance (as a fugue of two melodies with different rhythm in a musical piece).[4]

Indeed, both happen within the same horizon, namely the spatiotemporal horizon of corporality in general. However, some are internal culminations or, as Aristotle says, self-fulfillments (*entelecheiai*, *energeiai*) of the complete bodily system as a living unity; the other are partial states that are consistently spatiotemporally distinguishable, and relate as "ink" to the readable syllables of life episodes.

The culmination would not be possible if the foundation for it were not given, the "being alive" of the whole individual. According to Aristotle, for all psychic events or psychic states, we should keep in mind the difference between "primary" and culminating *entelecheia*, between the foundation-giving actualization or activity and the summit-building one. The first one, that is, the primary *entelecheia* of the whole system is, according to Aristotle, the defining concept of the soul; the second one, the *entelecheia* that is brought to culmination, is the common expression for all kinds of soul functions, which Aristotle sometimes also calls *praxis* or *chrêsis* [use made of something] (cf. *De An.* II 4, 415a19 und *PA* I 5, 645b14-22; *Metaph.* IX 8, 1050a23-b2.). The relation between soul and psychic state is not that between an underlying subject and a property, but rather the relation between basic disposition and performance or between basic form and developed culmination—like the distinction between the base and the maximum of a mathematical function (e.g. a parabola). While *primary entelecheia* is the self-propagating generic trunk of life-activity in each individual (to be identified with what Aristotle calls "the soul" of a living being),[5] the culminating activities or particular life episodes are like different graftings on that trunk which bear the mental and psychic states as their varying features.

I think that this Aristotelian image of psychic phenomena can be transferred to contemporary debates without infringing upon principles of modern scientific thought. Psychic states can never be identified with somatic states and processes that are determined on bodily parts and are added up out of bodily parts into particular regional patterns of activity. Rather, they are embedded without exception in life episodes that are biographically relevant, that is, they characterize the life-sustaining and life-configuring behavior of the whole organism.

Within these life episodes there are individual psychic phenomena that can best be described as "operative states" of the whole system that accumulate in operations of the living being in question (for instance motor, aesthetic, linguistic, cognitive operations). The sameness and variability of psychic states involves the sameness

[3] I have described this in more detail as an Aristotelian model of a possible solution to the body-soul problem from a contemporary perspective in: 2006b, pp. 85–106.

[4] Cf. for this concept of a weak dualism in the same horizon of the corporeal in general, Buchheim 2006a, pp. 38–49.

[5] Cf. *De An.* II 1, 412^{a}27 f.: "That's why the soul is a *primary entelecheia* of a natural body having life potentially in it." (transl. slightly modified T.B.).

and variability of the operations—this is, in any case, my thesis. The characteristics of psychic states indicate the kind of life episodes to which they belong and the kind of operations that can be achieved through them. If you can't hold your balance, you can't ride a bike; if you can't see, you can't paint; if you can't add, you can't multiply or calculate a complicated equation; if you can't hear the beat, you can't dance, etc.

It is always certain operative complete states with a certain psychophysical profile that enable incorporation in certain activities and operative sequences, and with them the *progress* of corresponding actions. An action or operation fails if the operative states do not possess the corresponding psychophysical profile that represents the key to the progress of the action. We will never cross the thresholds of certain actions; others are crossed only after a long time and practice; while still others are crossed after only a short time or even from birth. The culminations of our life episodes are accordingly simple or demanding, rare or commonplace.

I have already mentioned the expression by which I want to refer to the pairing relationship between psychic-biographical being and simultaneous bodily correlate (be it a state of the brain or of circulation or of the skin, or other somatic organ states): corporeal states are *symptoms* of the complete situation ('life situation') of a living being characterized psychically. All psychic states and processes are (according to the first argument) characteristics of a life situation in which the whole living being finds itself; all somatic states (according to the second argument) are only certain symptoms of this situation, and together they do not make up the psychic state. We also do not know exactly which groups of symptoms accompany which psychic events and clearly such correspondences do not always exactly recur.[6] Unreasonably, we tend to consider just neuro-physiological symptoms as complete correlates of psychic or mental states. However, this is barely justifiable.

It is possible, of course, that particular patterns of neuronal stimulation are typical to a greater extent for particular psychic states or mental performances, and so the former are symptoms that are necessarily paired with the latter, while other somatic symptoms seem interchangeable. On the other hand, much research on brain injuries shows that very different areas and therefore also different neuro-symptoms are required for the same psychic functions. The often mentioned plasticity of the brain indicates that there isn't a special set of neuro-symptoms that is exclusively correlated with a psychic state or even identified with it. Symptoms remain a concomitant phenomenon and they are not themselves the whole matter of psychic life.

[6]Aristotle also pointed out this curiosity: **[6]** "It seems that all the affections of soul (*ta tês psychês pathê*) involve a body – passion, gentleness, fear, pity, courage, joy, loving, and hating; in all these there is a concurrent affection of the body. In support of this we may point to the fact that, while sometimes on the occasion of violent and striking occurrences (*pathêmata*) there is no excitement or fear felt, on others faint and feeble stimulations produce these emotions, viz. when the body is already in a state of tension resembling its condition when we are angry. Here is a still clearer case: in the absence of any external cause of terror we find ourselves experiencing the feelings of a man in terror." (*de An.* I 1, 403a16-24).

5.3 An Example of the Causality of the Psychic as Such

The false limitation, in my opinion, of the bodily symptoms of the psychic to the so-called "neuronal correlate" has led some to suppose that psychic states or mental performances are mere epiphenomena or causally irrelevant appendices of neuronal processes in our brain—a sort of superstructure or a shadow of what happens in the brain and makes us feel this or think that, etc. I am convinced that this is a one-sided reversal of the actual dependencies. It often seems the other way around, that specific symptoms—also neuronal—only follow a certain life situation and its psychic characteristics. When we don't even notice that our blood pressure is too high or that this or that brain function is impaired, then nothing changes in the established course of total behavior. It is only once we become aware of certain signals that we can see that also certain neuronal symptoms change. A very interesting experiment that Jose M. Carmena and Miguel A.L. Nicolelis made some years ago at Duke University documents in an impressive way what I want to claim here (see Carmena et al. 2003, pp. 193–208).

The team of Nicolelis und Carmena studied the behavior of monkey subjects operating robot prostheses through so-called Brain-Machine Interfaces. The question they wished to explore was which neuronal populations and states of stimulation underlie fine-motor operations of the arms or prosthetic arms. To this end, they let the monkeys play a computer game that they controlled with a joystick. The task was to hit, as soon and effectively as possible, a blinking point that appeared on the screen. During the game, the monkeys' brain activity in certain brain areas was recorded, with a relatively fine definition for those neuronal populations that were supposed to elicit the control movements of the arms.

The monkeys learned to play the game pretty quickly and they liked to play it. At the same time the brain signals, additionally conditioned by a simple learning program, were applied to control a robot arm that began to make movements that were similar to the monkey's. In order to pinpoint the neuronal signals relevant to motor control, the contacts between the monkey's joystick and "his" computer were interrupted after a certain time and the computer was instead fed signals directly from the monkey's *brain*. That is, the monkey's arms were not in fact controlling the game anymore, but the brain directly. At first this arrangement led to a serious decrease in the game performance of all monkeys; but following a significant change in behavior and a reorganization of the neuro-active symptoms, game performance rose again to almost original levels. In an attempt to compensate for the sudden decrease in performance, the monkey would first make protruding movements with its arms that were too big. Gradually, after this strategy failed, the relevant neuronal populations and their stimulation curves changed until finally this neuro-activity led to results that were similar to previous levels, but that were achieved in a different and newly organized way. The monkeys realized pretty quickly that the movement of their arms was causally irrelevant and henceforth controlled the computer game without arm movement, through brain activity alone.

The development of this experiment clearly shows what kind of phenomenon seems to be causally subordinate to the other: the decrease in game performance changes the life situation of the monkeys that is characterized, let's say, through disappointment and anger about the results being suddenly worse. The anger produces an effort to improve the situation again. This effort leads to significant reorganizations of bodily symptoms: for instance, at first, to protruding arm movements with the control pole. Since this does not help, the neuronal populations in the monkey's brain, which by now are firing chaotically, are selected in a different way than before: those that lead to signals that improve the game performance are *favored*, as they bestow renewed success in the operation sequence of the game; others are repressed and ebb away due to irrelevance. In this way, the monkey learns to control the computer game through slightly reorganized neuronal activity—that is, directly with the brain. What is important for us is that the reorganization of neuronal symptoms in the operation sequence of the game *follows* the biographically affected life circumstances with their embedded psychic states and not the other way around.

Because the monkey gets angry and attempts to return to the earlier successful state, neuronal activity is selected differently and the best variables for game success are favored. In general terms, we cross the threshold of the corresponding next step of a biographical operation by reorganizing the bodily symptoms that accompany its performance. When we balance an egg on a spoon and run to the finish line, we reach the next step and the finish line only if the somatic symptoms are constantly reorganized according to the success of the operation. Now this seems to be valid not only for balancing acts and arm movements that we have learned to control voluntarily and consciously, but also for neuronal populations and stimulation patterns in our brain, as the experiment above teaches. They also develop in relation to their conduciveness for following the steps of the operation at issue; we make variations chaotically and serendipitously until we advance, thanks to an appropriate neural population and activity, to the corresponding next threshold of the desired operation.

Our operative abilities of thought, for instance in calculating or reading, could certainly be built this way. And since the brain has such plastic properties, operations achieved successfully once or several times are retained as a good path for the fruitful progress of our life.

5.4 A Proposal for a General Model: Psychophysical Causality Through "Favoring"

In this way it seems possible to describe a general model for the causality of biographical episodes and the psychic states embedded in them: the organic body of a living being is not subject to a unitary causal succession from one moment to the next; rather it builds a structure of functional systems that are relatively strongly demarcated from each other but overlap and are thus capable of coordination. These systems in turn break down into multiple subordinated causal connections. For this reason, the bodily symptoms of different biographical episodes and the psychic and mental states embedded in them are often dispersed throughout the whole body, they build patterns and relations that are not immediately connected to each other in a causal connection relevant to the complete behavior but are rather symptomatic *expressions* of the behavior or operative state of the whole living organism, as was described above by means of Aristotle's arguments.

This is why the sum of somatic symptoms, including neuronal correlates, is available according to the life situation and the biographically adequate way of behavior through which the organism is maneuvering. When we learn something, for instance, we expose our body to a situation for which it is appropriate to re-organize a particular subset of the somatic symptoms of our operation. For instance, we repeat a certain foreign word or sound until we can articulate it fluently and correctly. Or we try to stay on the bike saddle until it becomes easier to find balance at higher speeds. We create circumstances, then, for one another and also just for ourselves, in which particular symptoms, including neuronal correlates, are selectively favored, in order to advance to the next corresponding threshold of our operations. When we are practicing something, we remain before this threshold, searching and chaotically varying these symptoms, until the threshold is crossed and the operation progresses.

As mentioned before, what is unacceptable is the view that these successful operative states (as formulated in Aristotelian manner, the "culminations" or *entel-echeiai* of being alive) are *identical* with the corresponding somatic symptoms, neuronal patterns, etc., because in that case we would have to renounce the truth of the affirmation that psychic and mental states and sequences *as such* and in their non-somatic characteristics could be *causal* for the somatic continuation of our existence.

We cultivate thought because its symptomatic expression brings enormous advantages and improvements to our behavior and thus to the somatic profile of the course of our life. These advantages are due to our *thinking*, not to the neuronal firings that, without attention to operative rules of thought, could take a completely different direction in each person. Thought as such has certain characteristics, as is probably clear to everyone, that no purely somatic process or set of processes can possess. Some examples of these characteristics are:

- Intentionality (to mean something, significance)
- Subjectivity (experience quality; first-person perspective)
- Reflexivity (self-transparency; awareness that I am the one who is thinking)
- Integration of foreign perspectives (empathy, communication, speech-character)
- Negative attitude (in logical or practical contexts)
- Truth-orientation (thoughts aim at truth)
- Normativity (we take pertinent norms into account in our actions)

These and other characteristics must be able to leave their causal footprint in our corporeal existence.

Therefore, it is *insufficient* to say that mental dispositions *are* mere neuronal states (identity theory; see e.g. Beckermann in Pauen and Stephan, pp. 122–147), that mental dispositions supervene on neuronal ones (without their own causal relevance), or that they have causal relevance due to their identity with neuronal states (anomalous monism; see Davidson 1980, pp. 113–116). In all these models the characteristics of thought I mentioned are *causally depotentiated*. The causal explanation for what happens is not thought, but neuronal states. However, in the experiment described above, it was clearly the monkey's *desire* to improve its game performance that was the cause for the reorganization of its brain activity. The monkey finds itself in a peculiar life situation that affects not only the state of its brain, but also its complete disposition as an agent with certain interests and experiences (which certainly possess intentional characteristics). This situation *favors*, as we said, the production of particular neuronal states and neglects others. A brain always produces whole *populations* of related micro-states that differ from each other in operationally relevant ways. Such variations can be accentuated because of their being more favorable for the progress of ongoing life-episodes or they can be eliminated through neglect.

The concept of favor or serendipity is meaningful and important in any case where change in a comprehensive context is systematically paired with change in the individual components or symptoms of the context. We find many examples of a causal pairing like that in economics and other 'trend' phenomena like the weather, fashion and style. In all these cases a particular change in the component involved can favor or compromise another particular change in the comprehensive context. And vice versa, a particular change in the comprehensive context can favor or compromise a particular change in its components. Also related in this way are the comprehensive life situations in which an organism finds itself and the symptoms of individual body parts that accumulate over certain operative states of the whole organism.

Thus, in my view, a model of psychophysical causal connection through favoring could look something like this:

(1) We learn by maneuvering each other and ourselves into certain life situations and so creating circumstances that are adequate for particular productions and not for others, and which thereby modify our life situation. We are in turn able to incorporate these favored productions into similar situations as enhancing characteristics; that is, we can then maneuver ourselves again into these situations in order to become more secure through further favoring of the corresponding states, that is, in order to practice better performance ("favoring spiral").

(2) Such a favoring spiral is the grounding basis of operation. An operation is distinguished from a sequence of events because its individual phases are united by operative connections rather than immediate causation. That is, they are connected through a sequence or development of life situations through which favored bodily states (including neuronal states) are produced one after the other. Balancing a bowl full of soup so that it can be brought to the table is an example of such an operation. The muscular contractions and angles of the bodily components depend on sensations of how much the level fluctuates. If

you don't master the operation, the waves will inevitably build up because you allow adverse tensions free play—you have selected them wrongly.

(3) Operation attempts slowly create passages and bridges of capacities or, even better, abilities. This means more or less long segments in which the sequence of life situations remains on track for the operation and does not slip out of hand as it does for the beginner soup-balancer. Favoring is already preparing for the situations that will follow the current one. That is why there is disappointment when someone does not master operations flexibly enough.

(4) The fourth step consists in the fact that we almost always avoid disappointment, that is, we bring our operations to an end by adapting to the special situation through which we must "navigate" the sequence of life situations. It is still important for all human operations to create particular adequate spaces in which operations can be successful, especially if they are very complex. Also, for the same reason, we do much to help each other, as at the beginning of learning. So in general it is possible that we behave in a particular way in a causal sense *on account of* the specific characteristics of our mental dispositions like intentionality, subjectivity, contextuality, reflexivity and intersubjectivity of thoughts and volitions, etc.

If this model of mental causality is on the right track, then theories that identify the mental with what is physically describable are clearly false: concepts of supervenience as well as anomalous monism without psychophysical causality are insufficient. For this model demands an authentic causality of mental states, and while such states would indeed be states *of* physical-material beings, they cannot be themselves again purely somatic, i.e. physically describable and explainable states of such beings. If instead, in the described way, causally relevant mental states of certain material beings taken as a whole existed which aren't to identify with their correlated corporeal symptoms, then an essential mark of the mental would be especially highlighted; namely that we can, do, and must enable *each other* to make such mental states an ever more important aspect of our lives. Since we must be originally "maneuvered" into life situations, this seems to me a key perspective for social neuroscience (see Cacioppo et al. 2002).

References

Aristotle. 1984. *The complete works of Aristotle. The revised Oxford translation, I*, ed. J. Barnes. Princeton: Princeton University Press.

Beckermann, A. Die reduktive Erklärbarkeit des phänomenalen Bewusstseins - C.D. Broad zur Erklärungslücke. In Pauen M. and A. Stephan eds.

Buchheim, Th. 2006a. *Unser Verlangen nach Freiheit*, 37–66. Hamburg: Meiner. bes. S.

Buchheim, Th. 2006b. *Sômatikê energeia* – ein aktualisierter Vorschlag des Aristoteles zur Lösung des Leib-Seele-Problems. In *Das Leib-Seele-Problem. Antwortversuche aus medizinisch-naturwissenschaftlicher, philosophischer und theologischer Sicht*, ed. F. Hermanni and Th. Buchheim. München: Fink.

Buchheim, Th. 2012. Neuronenfeuer und seelische Tat. Ein neo-aristotelischer Vorschlag zum Verständnis mentaler Kausalität. *Philosophisches Jahrbuch* 119: 330–346.

Buchheim, Th. 2013. Ein neo-aristotelischer Vorschlag zum Verständnis mentaler Kausalität. Eine Replik. *Philosophisches Jahrbuch* 120: 371–393.

Cacioppo, J.T., S.E. Taylor, et al. 2002. *Foundations in social neuroscience*. Cambridge, MA: MIT Press.

Carmena, J.M., A.L. Nicolelis, et al. 2003. Learning to control a brain-machine interface for reaching and grasping by primates. *PloS Biology I* 1(2): 193–208. http://biology.plosjournals.org.

Davidson, D. 1980. Mental events. In *Actions and events*. Oxford: Clarendon.

Hacker, P., and M. Bennett. 2003. *The philosophical foundations of neuroscience*. Oxford: Blackwell.

Kurthen, M. 2006. Der Augenblick des Bewusstweins und die lange Zeit des Gehirns. Über den möglichen Beitrag der Kognitiven Neurowissenschaft zur Lösung des Gehirn-Geist-Problems. In *Das Leib-Seele-Problem. Antwortversuche aus medizinisch-naturwissenschaftlicher, philosophischer und theologischer Sicht*, ed. F. Hermanni and Th. Buchheim. München: Fink.

Strawson, P.F. 1959. *Individuals. An essay in descriptive metaphysics*. London/New York: Methuen, Reprint.

Stump, E. 1999. Dust, determinism, and Frankfurt: A reply to Goetz. *Faith and Philosophy* 16: 413–422.

Chapter 6
Does the Principle of Causal Closure Account for Natural Teleology?

Miguel García-Valdecasas

The principle of causal closure, otherwise known as 'the canonical argument for physicalism' (Papineau 2002, p. 17) has purportedly intensified the relation between physics and metaphysics by showing how both sciences can advance a more profound understanding of physical phenomena. In its basic formula, causal closure says that every physical event has a physical cause, and that the physical world is self-sufficient and causally complete (Papineau 2002, p. 234). Although some physicalists may deny that physicalism buttresses materialism—the view that everything is basically matter and that only the material world is irreducibly real—at the end of the day most physicalists hold that everything is either physical or supervenes on the physical (Stoljar 2009), so that ultimately, the physical is essentially all that there is. The material world is all that physics and metaphysics must deal with to give us a complete view of nature.

In the view of causal closure, the existence of non-physical objects can only be speculated about or wildly hypothesised. It is true that philosophy has traditionally endeavoured to gain access to transcendental, non-physical objects, which central historical figures of philosophy considered essential to any lasting explanation of the actual world. Yet in the context of physicalism, theories that countenance and try to categorise non-physical objects, if cogent, remain at best unrefined or inadequate. In the view of causal closure, there are basically two ways of making sense of non-physical objects. If non-physical objects have causal efficacy, physics should eventually reckon with them; we are simply required to wait until physics explains them. If they lack causal efficacy, there is nothing that physics can say about non-physical objects. This implies that, in effect, every explanation of events which *prima facie* falls outside the scope of physics, or of the hard sciences in general, may never be

M. García-Valdecasas (✉)
Mind-Brain Group, Institute for Culture and Society (ICS), University of Navarra, 31009 Pamplona, Spain
e-mail: garciaval@unav.es

M. García-Valdecasas et al. (eds.), *Biology and Subjectivity*, Historical-Analytical Studies on Nature, Mind and Action 2,
DOI 10.1007/978-3-319-30502-8_6

more than provisional. One can reasonably expect better explanations that will supersede them over time. In principle, we are entitled to question whether an explanation of natural events that falls outside the scope of physics can be a valid or ultimate explanation of these events. Since the proponents of causal closure hold that the true explanation of what there is the physical explanation, I examine how this principle engages the existence of non-mechanistic causes. I argue that these causes exist, that they require explanations, that physicalist causal closure cannot provide such explanations and that, in consequence, the physicalist worldview of closure demands too much of reason. The endorsement of physicalist causal closure renders many natural phenomena unintelligible—biological phenomena in particular. Accordingly, the world is not as simple as it looks.

The argument has three sections. In the first section, I briefly discuss the contents of the principle of causal closure and some of its consequences. I argue that, paradoxically, the principle of causal closure shares central assumptions with the very theory that is supposed to reject: Descartes' dualism. In the second section, I contrast the principle of causal closure with Aristotle's world of natural ends, which under the premises of this principle can only be accommodated as a non-physical and inherently extrinsic force. However, a number of natural facts point to the existence of natural regularities and goal-directed behaviour. I argue that science cannot be effectively done without positing crucial teleological assumptions that guide the scientific analysis toward what is essential for explanation. Finally, inspired by Aristotle's view of ends, the third section articulates a principle that prevents the reduction of forms ands ends to more basic or constitutive natural elements, and discusses the compatibility of this principle with the scientific method. My hope is to show that, by developing a more profound understanding of natural forces we can better appreciate the enduring value of Aristotle's teleology.

6.1 On the Meaning and Significance of 'Causal Closure'

Let us start by stressing that although causal closure is concerned with physics, it is a metaphysical rather than a physical theory. It is clear that physics may have little or nothing to say about what the physical world is, as opposed to how it physically behaves in the contexts that are relevant to physics. Accordingly, in its traditional description causal closure establishes that no physical event has a cause outside the physical domain. In other words, the theory holds that every physical event must have a physical cause. More precisely, Kim argues that 'if we trace the causal ancestry of a physical event, we need never go outside the physical domain' (1993, p. 280). Or in Papineau's most recent formulation, 'the causal closure of physics entails that every physical effect has a sufficient physical cause' (2009, p. 53).

In parallel, Papineau states that 'the causal closure of physics is solely a claim about how things go within physics itself. It does not say that everything is physical' (2009, p. 54). In his view, physics is a self-sufficient explanatory realm, and so are sciences that base their knowledge on physics. In physics, the explanation of sea tides does not need to invoke biology—physics can satisfactorily deal with sea tides; in biology, the explanation of the ventricles of the heart does not need to invoke psychology—biology can satisfactorily deal with ventricles of the heart, and so on. Of course, the centrality of physics does not imply that every possible existing object must be physical. What is more, causal closure cannot preclude the existence of non-physical entities. 'It is entirely consistent with the causal closure of physics that there should be self-sufficient realms that operate quite independently of physical goings-on' (2009, p. 54). Likewise, in Lewis' view, the hypothesis that there is a unified body of scientific theories

> does not rule out the existence of nonphysical phenomena (…) It only denies that we need ever explain physical phenomena by non-physical ones (…) All manner of non-physical phenomena may coexist with them, even to the extent of sharing the same space-time, provided only that the non-physical phenomena are entirely inefficacious with respect to the physical phenomena (1966, p. 23).

Of course, one could accept Lewis's notion of a physical domain that is somehow cordoned off from non-physical domains. But the way in which the physical domain becomes the central domain is not thereby explained. In Papineau's case, the centrality of the physical domain as the realm of causal efficacy is speculated about but never set out in much detail. So, even if we posit the existence of non-physical domains, to the extent that causal closure undermines the explanatory value of theories concerning these domains, these are given a marginal existence and reduced to provisional status. What is more, Papineau notes that causal closure 'does give rise to a powerful argument for reducing many prima facie non-physical realms to physics: for it indicates that anything that has causal impact on the physical realm must itself be physical' (2009, p. 54).

For instance, the principle stipulates that if any event in the biological and mental domain has physical effects, this event must have a physical cause. Regardless of other possible effects of this event, if we know that at least one of its effects is physical, its cause will also unquestionably be physical. Papineau does not say whether this cause is necessary or sufficient, but contextual cues indicate that the cause must be a sufficient cause in its own right. Therefore, following Papineau, possible non-physical causes (if real) will be so 'prima facie,' because if non-physical causes have physical effects, physics cannot satisfactorily account for such an explanation. Thus, in asserting a preference for physical explanations over other possible kinds of explanation, Papineau hopes to strengthen and further develop the relation between science and philosophy. As he puts it: 'The causal closure of physics promises to bring philosophy into close contact with science at a more detailed level' (2009, p. 60), that is, at the level of physical explanations.

Nowhere in this account do we see the evidence on which causal closure is grounded. We are told that the world exhibits a closed and self-contained structure whose centre is the physical domain. Even events that may not be fully physical will be more or less physical epiphenomena—depending presumably on how much ground we are willing to cede to reductive physicalism—in a world of fundamental physical laws.

I do not intend to discuss causal closure here on its own terms, even if—in my view—those terms are highly metaphorical and assign a role to physics that many physicists find idealistic or fanciful. It is questionable whether the physical explanation is the best possible explanation when some of its branches such as quantum mechanics seem to be mired in a slew of quandaries about causality, the nature of quantum phenomena and the place of the observer in it. Thus, instead of discussing the plausibility of causal closure as such, I will contrast this theory with the perspective of Aristotle's philosophy of nature, where biological phenomena rest on complex irreducible dimensions like teleology. This comparison is motivated by the fact that, although philosophy has traditionally rejected Descartes' dualism for its watertight separation of physical and non-physical realities, mainstream physicalism has sponsored a view that seems to share a number of Descartes' central assumptions.

How does physicalism share central assumptions with Descartes? Descartes contended that the extended, physical world is an autonomous realm of matter, motion and energy exchanges. The laws of physics and other natural sciences satisfactorily determine the movements and trajectories of these objects. Its laws are complete, in the sense that they do not need to invoke non-physical laws in order to offer a complete explanation of the objects of the physical realm.

Descartes identified a distinct, higher realm of non-physical entities that do not display motion nor have extended parts in the manner of physical entities. If we take Descartes' view of the autonomy of physics in its domain and simply dismiss the non-physical domain, we have something that is fairly similar to mainstream physicalism. Descartes' watertight division of separate ontological domains and the ontology of causal closure are hence strikingly similar. For one thing, Lewis' ontology depicts the possible coexistence of events that are causally inefficient with events that are causally efficient. For his part, Papineau presents us with a multiple-tier view of reality that is not significantly different from Descartes' substance dualism, except for its attempt to conceive all phenomena, including life and reason, as physical. Papineau emphasises the importance of keeping physical objects within the domain of physics, and to leave open the possibility of coexisting non-physical phenomena. But as physics makes more progress on closer inspection the latter may turn out to be physical phenomena. As he describes it: 'causal closure (…) argues that many prima facie non-physical occurrences (…) must themselves in fact be physical' (2009, p. 55), so that we can conceive of a reduction or collapse of domains. Less straightforward options such as inter-realms causation cannot be considered, for 'otherwise it is hard to see how [what lies beyond physics] could have any physical effects' (2009, p. 55).

Papineau argues that anything that brings about a physical change must have a physical cause. Thus, when for instance, any fibre in one's central nervous system is stimulated as a result of a desire to raise one's left hand, this desire must be a physical or biological occurrence that finds its explanation in physics or biology. Causal closure remains agnostic or uncommitted about whether this desire also figures in some possible explanation to be found in higher realms of reality such as psychology or sociology. Causal closure has nothing to say about those explanations, regardless of their plausibility. To play it safe, causal closure simply sticks with the belief that the stimulation of fibres in the nervous system is all that is needed to articulate a satisfactory explanation of desire.

The similarities between Descartes' philosophy and physicalism are noteworthy. Both theories share the conviction that every realm is complete, separate and self-sufficient. Accordingly, they also share the consequence that inter-realm relation is problematic—Descartes' pineal gland being no more than an *ad hoc* solution. They differ, however, in the significance given to the physical realm. While physicalism is built upon the physical realm and sees every other realm as supervening on the physical, Descartes believes the physical realm to be of lesser explanatory value than the non-physical realm, where philosophy has historically gained a better foothold.

Be that as it may, Aristotle's ontology is at odds both with Descartes' theory and with mainstream physicalism, the parent theory of causal closure. While closure derives its ontology from mainstream physicalism, physicalism is a form of materialism. Materialism is an unbroken philosophical tradition that stretches back to the atomistic view of Empedocles and Democritus. In his own times, Aristotle came to know their views and extensively discussed some of their central assumptions. Aristotle argued that Empedocles and Democritus' view of nature as a mere combination of matter and mechanical forces involves a gaping deficiency, as nature cannot be simply understood as the interaction of atoms. More will be said about this in the last section.

Aristotle's ontology is also distinguished for its monism. He does not divide the world into multiple possible or existing realms. There is only one realm: the real world. But in contrast with physicalism, in Aristotle's ontology the laws of physics can only explain a part of the intrinsic dynamics of nature, thereby preventing the causal confinement of this world to the basic laws of physics. He contemplates the action of mechanical and non-mechanical causes together. To say that non-mechanical forces are 'non-physical' would be inaccurate, but to say that they are 'physical' does not imply that they are intrinsically material. While non-mechanical forces are physical, they are not so in the way of matter. Certainly, to show that such heterogeneous causes can plausibly interact, Aristotle must convince us of the existence of such non-mechanical causes. And if we can articulate a plausible argument for their existence, their outright dismissal may entail the loss of a significant part of nature and impair our epistemic analysis. To avoid this prospect, let us review Aristotle's world of natural ends.

6.2 Aristotle's World of Natural Ends

Aristotle's natural world includes a variety of substances from the smallest bits of matter to God, the first unmoved mover and the most perfect substance. This world has two separate regions—the supralunar and the sublunar—that do not mechanically interact. But this fact does not mean that the two regions are radically separated. There is in fact metaphysical interaction between them. For example, Aristotle suggests that God, the first mover, acts as an attractor for all other substances, not just those located in the supralunar world, being the ultimate cause of the movement of every substance. Acting as the final cause of the whole universe, the first mover manages to attract towards itself other substances in the supralunar and sublunar regions. If this is not an oxymoron and the first mover actually succeeds in attracting all sublunar substances, this relation cannot be explained by efficient cause, namely, by the pushing and pulling of the first mover. Given that the two regions cannot interact mechanically, their relation can only be explained teleologically. Hence, it seems that rather than impregnable domains of the kind posited by Descartes, Aristotle's two regions make up a unitary and well-integrated world.

Is the world of causal closure equally integrated? Let us examine this point. We can say that the world of causal closure is one because it purports to explain all existing objects from the perspective of physics. Although the claim of causal closure is 'solely a claim about how things go within physics', the context in which the principle is formulated assumes the sufficiency of physical explanation. In its wake, high-order levels of reality submit to the lower levels that are the focus of scientific analysis. Thus, by recommending a bottom-up explanation of reality, causal closure seems to frame a compelling argument to characterise all causes as physical and assume that the perspective of physics is good enough to guarantee the overall unity of nature.

Hence, closure describes a highly integrated world, probably more so than Aristotle's universe where the perspective of physics does not play a comparable role. Still, in contrast with Aristotle's universe, causal closure builds its world around the material cause. By 'physical causes' Papineau means the matter and energy of modern physics, a world where matter is ubiquitous. If any object O can be found in this world, O must be material. Even if an object P appears to have properties that do not fit within physical categories, the overriding assumption of the philosopher should be that P will be so only prima facie. Papineau is convinced that if P has effects on the physical domain, on closer scrutiny its material constituents will undoubtedly emerge. In this way, the basic assumption of closure is that either every object is a material object or that it owes its existence to some material object.

I will try to show that the cost of this strategy is high for both science and philosophy. Causal closure regards everything as causally material, or if open to the perspective of many worlds, as having independent material effects. Now we may wonder: is causal closure a principle which science and philosophy can recommend?

To discuss this issue, I will review Aristotle's concept of teleological causation. The discovery of ends seems to have afforded Aristotle a vivid and unprecedented account of varied phenomena, whether biological, ontological or ethical. In this context, many have found teleology to be the most distinctive mark of Aristotle's philosophy (Johnson 2005, p. 1). At the same time, teleological causation has often been a stumbling block for contemporary scientists, who repeatedly pointed to the flaws of metaphysical or mentalistic notions like teleology. For instance, teleology has been accused of preventing the understanding of evolution by natural selection, a process that is incompatible with Aristotle's belief that species are unchanging and eternal. Despite this, Aristotle's teleological arguments derive from natural observation, and he regarded teleology the best possible explanation of natural regularities. In his view, natural regularities or events that repeat themselves 'for the most part' manifest an intrinsic natural or hypothetical necessity, such as the necessity of some mammals to grow sharp teeth in front to bite and tear food. In his view, the growth and development of teeth in mammals is not blind or random, but is instead guided by a biological principle that is intrinsic to the nature of every substance. Knowledge of this principle enables us to understand why the growth and development of teeth obtains with such frequency. Aristotle repeatedly claims that 'everything that comes to be moves towards a principle, i.e. an end' (*Met.* IX 8, 1050a 7–8). In a slightly different formulation, he claims that 'nature does nothing in vain. For all things that exist by nature are means to an end, or will be concomitants of means to an end' (*De An.* III 12, 434a 30–32). And again: 'nature, like thought, always does whatever it does for the sake of something' (*De An.* II 4, 415b 15–16).

Teleology is best understood as a cause that works in conjunction with other natural causes. In particular, ends owe much of their existence to formal causes. Although ends and forms are distinct causes, Aristotle often conflates them in his *Physics*, noting that the formal and final cause are often *(pollákis)* the same (II 7, 198a 24–27). The formal cause presents matter as an informed reality, namely, as distinct and identifiable from objects of a different nature. In a similar way, following Aristotle we might say that knowledge of the characteristic end of some object provides a richer and more coherent understanding of its movement. Aristotle often portrays the natural end to which natural things move as the reason (*logos*) or heuristic principle by which they move. So the end of every substance articulates the reason why this substance moves and, ultimately, why it is structured in such a way as to perform a specific function. In the case of mammals, the need to bite and tear food explains why they grow sharp teeth. In other cases, Aristotle uses the final cause to explain what makes instruments like an axe the kind of instrument that it is, and what makes an organ like the eye the kind of organ that it is. We could roughly define these things functionally, that is, by saying that an axe is an instrument whose elements are arranged in a certain way for chopping wood; the eye, in turn, is the organ of sight. By invoking the kind of activity that is most characteristic of both, we identify a defining reason for the way in which they are. Aristotle illustrates the idea using slightly different words in the following passage:

> That for the sake of which a thing is, is its principle and the becoming is for the sake of the end, and it is for the sake of this that the potentiality is acquired. For animals do not see in order that they may have sight, but they have sight in order to see (*Met.* IX 8, 1050a 8–11).

In Aristotle's view, both the axe and the eye are instruments that exist for some purpose. Such 'principle' or explanatory *logos* explains the reason why these instruments have their both physical and characteristic make-up and their capacity. This phenomenon has traditionally been explained by the so-called 'hypothetical necessity', which Aristotle lays out in *Physics* II 9. Hypothetical necessity articulates that we must presume the existence of ends as a working hypothesis in order to explain the existence of material necessity in nature. In other words, without the hypothetical necessity introduced by the *explanandum*, we would have a hard time understanding what kind of function an organ such as an eye might play. Hypothetical necessity plays a crucial role in every scientific account. For instance, today we explain parts of the eye like the cornea, the iris or the fovea as parts of an organ that is adapted to seeing. In so doing, we stipulate some hypothetical necessity about the eye as an organ that fulfils a specific role. By assuming that the eye does something, we acquire a notion of what seeing is. But what if the vagaries of natural selection had taken a rather different evolutionary path, one that never gave rise to eyes? We might easily conceive of a world of varied creatures that adapted to their environment and prevailed independently or jointly without eyes. On this assumption, let us conceive that a specific animal of this world spontaneously develops eyes similar to ours but cannot make effective use of them. Given that the emergence of eyes in this animal could not be associated with any imaginable function, hypothetical necessity would unquestionably fail to explain the emergence of eyes. What justifies this assumption? We discover and understand the capacities of animals on the basis of some prior knowledge of their *logos* or ultimate reason, whereas knowledge of the material necessities of animals may arrive at a later stage. Unfortunately, we know very little about the way in which we grasp this *logos* as opposed to the way in which we grasp more evident causes such as the efficient cause. But although the way in which we apprehend *logoi* may be uncertain, the fact is that most objects would remain unexplained without the assumption of hypothetical necessity.

To understand how nature articulates ends, I will briefly sketch the role of the other fundamental natural causes, namely, the material and formal cause. In Aristotle's view, form and matter are co-principles or co-causes that are intimately united in material compositions, being closely involved in the generation of both living and non-living substances. Their stable union has been called the 'hylomorphic structure', a stable composition in which matter plays a basic role by supplying an underlying basis of change, and form plays a more sophisticated role by arranging the integration of parts in view of some end. As we will see next, the specification of the end of a substance represents a valuable defining mark for rational analysis, because the formal features give us crucial information as to why the substance is articulated in a given way. As Aristotle notes, 'the character of everything is determined by its end' (*Nic. Eth.* III 7, 1115b 22).

The thesis under consideration is that formal and final causes, though elusive, are necessary for explanation. To illustrate, let us consider a living organism. Organisms are made up of complex layers at every level, whether molecular, cellular or organic. In Aristotle's view, what ultimately explains these structures and their complex dynamics cannot be found at the molecular, cellular or organic levels, no matter how much we know about the processes that lead to their development. In Aristotle's view, such analysis would repeatedly beg the question of the reason *why* such physical entities are articulated as they are and work as they do. I noted earlier that we know what an eye is when we know about sight. In the same sense, we know a living organism when we have some prior knowledge about life: what a living substance is, as opposed to a non-living one, and the kind of ends that it pursues. To know that some substance is alive is to know what the substance is capable of, and to know what it is capable of is to know what it does and what distinguishes it from other similar substances. This kind of knowledge, which Aristotle attributes to the knowledge of some principle or some primary cause, cannot be called 'analytic' in the sense in which science is analytic, because it often precedes scientific analysis, guiding or contextualising it to provide a solid basis for understanding.

This implies that some essential knowledge of what characterises a natural species is prescientific. Consider the case of corals. Corals are maritime life forms that spread along the coast. Despite the hard stony surface that makes them resemble jagged rock, they are alive. The arrangement of their parts and their composition is adjusted to the pursuit of specific natural goals. Once we hypothesize that corals are not rocks but living things, the process of finding out how they work can get underway. That is, we can only know why coral *ultimately* develops in its characteristic way after we have posited that coral is a maritime life-form. In so doing, scientific analysis moves backwards until the working hypothesis that says that coral is a living substance is confirmed or disconfirmed by empirical data. The data are then informed by the working hypothesis until its content eventually becomes a reliable fact.

I know that this argument might be considered regressive for a number of reasons. As I cannot cover all of them, I will only address the circularity of the concept of nature. Since I am arguing that ends are no less intrinsic to nature than formal causes, to avoid circularity I should say something about Aristotle's concept of nature. Aristotle calls nature the first principle of movement, stating that it is 'the source from which the primary movement in each natural object is present in it in virtue of its own essence' (*Met.* V 4, 1014b 18–20). The nature of living beings is the genesis of its growth, nourishment and reproduction. Even though Aristotle can be rather explicit about ends, this definition, which includes 'natural objects' and motion, does not include any reference to natural ends. Still, in Aristotle's view, it is evident that nature as the source of motion is intended to account for the full extent of motion, not just for its beginning or its later development. And the full extent of motion necessarily includes its end. Above I quoted Aristotle as saying that 'nature does nothing in vain' (*De An.* III 12, 434a 30). Despite the fact that some movements appear to be random and still others are genuinely caused by chance, natural movement is not random or purposeless.

Aristotle has a compelling reason to argue that natural regularities such as the growth of plants are goal-directed. He expresses it in his debate with Empedocles, who had gained notoriety for a theory that ruled out internal ends from the movements of life forms. This exclusion affected plants, of which Empedocles allegedly claimed that their roots grow downwards 'by the natural tendency of earth to travel downwards and the upward branching by the similar natural tendency of fire to travel upwards' (*De An.* II 4, 416a 1–2). For Empedocles, the growth of roots and branches was hence driven by the thrust of their constitutive elements—or what I will also call 'element-potentials' in the next section for their development of the potentialities present in the constitutive elements of substances. But for Aristotle this explanation fails to realise the limited role that such natural elements play in growth. Instead, he invites us to consider what holds together contrasting elements like earth and fire, which, in his physics, are said to have contrary tendencies. He notes that if there is no counteracting force, the roots of the plant will be torn asunder (*De An.* II 4, 416a 6–9). In the absence of a high-order force, the natural tendency of the constituent elements of roots will destroy the plant. Empedocles is thus confusing a concomitant cause with the main cause (*De An.* II 4, 416a 14) that stabilises the plant as an organic unity. What keeps the different organs of some plants together and neutralises negative side-effects is their *psychê*, that is, the principle by which a substance is said to be alive and has an internal unity.

Thus, in identifying *psychê* as the main cause of life forms, Aristotle assumes that the natural elements of a plant have been co-opted by higher forces and are no longer free to move in the way in which they would if unconstrained. The plant has an inherent capacity to leverage external forces and integrate them for their benefit. Further, the capacity of life forms to perform such a change is pervasive, that is, it is at work in any living form. For instance, biology regularly explains the perceptual abilities of animals as crucial to understanding the role of their species in their native ecosystem, in which such animals turn out to become particularly successful because of that feature. Take the case of insects such as mosquitos. The heat emitted by mammals serves as a powerful attractor to many kinds of mosquitos. Mosquitos guide themselves through the darkness by sensing the infrared waves emitted by these mammals. They have also developed an acute sense of smell, which they combine with their sensitivity to heat, and possibly to pressure, to identify the precise location of mammals. Thus, once we know that the mosquito can track sources of heat in a nocturnal foray, the foray of the mosquito is no longer random. We understand what mosquitos pursue, and ultimately, what kind of role such attraction might play in their ecosystem because prior to that we assumed that mosquitos have perceptual abilities. In biological explanations, such a role is often believed to underlie the capacities of a living organism. This analysis invites us to look at the entire ecosystem as defining a varied set of goals for the adaptation of species and the kind of organs that are suited to such an adaptation. Again, we can illustrate this role if we consider the case of wild predators. Wild predators are said to keep the numbers of other groups of more actively reproductive animals in check, that is, within a range that strikes more or less sustainable balance between predators and prey. Despite the fact that predators have an overwhelming advantage over prey in terms of speed and strength, this advantage is compensated for by their low fertility rate, because predators cannot proliferate at the rate of their prey. In this way, we come to see how a self-regulated interspecies relation compensates for the apparent vulnerability of prey with other competitive advantages. Once again, the self-maintenance of the entire ecosystem is a relevant case of a kind of teleological argument that inspires biological research by providing a strong and effective hypothesis.

In the literature, this kind of teleology is often called 'interspecies teleology'. It may be described as the capacity of one natural being to appropriate and use the potential available to other natural beings for its own benefit (Leunissen 2010, p. 5). Interspecies teleology is characteristically addressed from the perspective of the animal that benefits from the existence of other animals, rather than from the perspective of the ecosystem as a whole, which in Aristotle's view does not constitute a substance and lacks a constitutive *psychê*. However, biological explanations often stress the role that self-regulated interspecies relations play in the development of most animal species. These animals require the existence of other species to adapt and carry out their internal possibilities. And if such relations in fact obtain, there may not be a plausible reason to deny that the features of the larger ecosystem are also important for understanding the rise and common acquisition of adaptive capacities that species develop over time.

Of course, this assumes the existence of interspecies teleology, or in more biological terms, the existence of an interspecies balance. And once again, such a balance may not be more than a working hypothesis until confirmed by empirical or statistical evidence. But whether it is highly plausible or confirmed by evidence, the fact is that biological theorists tend to pay little to no attention to the validity of the hypothesis of interspecies teleology as a scientific fact. Instead, biologists are commonly satisfied with the use of this hypothesis as an instrument that plays a decisive role in explaining the features and behavioural patterns of well-adapted and successful species.

If this argument is sound, let us compare the explanatory power of ends with that of the principle of closure. In a closed world we should expect physics or chemistry to exhaust the reasons concerning how the organs of predators seem to have a defining role in the interspecies balance. If we can do this without the perspective of other sciences, the explanation should be much better. Yet can this be a reasonable expectation? In other words, if we had to look into the organs of a predator to find out exactly how the interspecies relation works we would probably not know where to begin our analysis. For which parts or organs of predators incorporate the reason that, were predators to hunt disproportionally (too much or too little) or have a higher or lower fertility rate, they would critically upset the interspecies balance? What is usually expected from the analysis of an animal's organs is knowledge about properties that are beneficial to the animal itself, rather than to other neighbouring animals. Accordingly, it seems obvious that interspecies teleology cannot be grafted on any of the parts of the animal. For this reason, the rise of interspecies balance appears to be far too complex an issue for biology to expect a solution from an analysis of the organs of predators.

Without ends, the explanation of environmental features of substances as a whole becomes highly problematic. This is certainly unsurprising, because in the world of causal closure the natural arrangement of physical parts to some ends is reduced to a statistical fact. Whether the natural connection between the acorn and the oak tree is random or not, the principle of closure points to no specific reason why the former evolves into the latter except for the biological mechanisms of the acorn that facilitate its growth into an oak tree rather than a beech tree. If statistical arguments such as these were to pass as full explanations of natural motions, ends would ultimately amount to the fact that efficient causes bring about amazing successions of regularities by sheer coincidence. Aristotle criticises Empedocles for maintaining such a view:

> [Empedocles argued that] many of the characters presented by animals were merely the result of incidental occurrences during their development; for instance, that the backbone is as it is because it happened to be broken owing to the turning of the foetus in the womb (*PA* I 1, 640a 19–23).

For Aristotle, attributing the development of the backbone in the foetus to chance is tantamount to advancing no reason for it. Thus, if any feature of the natural world seems to explain why the development of the backbone in animals recurs with such frequency and precision, chance is not the reason. Nor are the biological mechanisms responsible for the development of the backbone. The question is not *how* the backbone is formed, but *why*. Thus, the ultimate reason for the frequency and precision of the development of the backbone becomes far too complex to be managed by a mechanistic theory. As a result, the why question and the ultimate understanding of phenomena like the development of the backbone in animals, growth and reproduction remain consistently elusive.

The reason why the world of closure cannot make sense of goal-directed behaviour and natural ends is that ends are analytically elusive. Extensive analysis of the physical properties of a substance does not unveil any ends in them. Final causes do not come to light at the microscopic or the macroscopic level as physical magnitudes such as size, temperature or speed, nor as a result of the combination of such magnitudes with other magnitudes. The elusiveness of ends, then, may be the outcome of our limited knowledge of substances, or of our attempt to focus only on natural features that match the scope of scientific analysis. Nevertheless, despite our limited knowledge of nature and the shortcomings of analysis, natural movements such as growth and reproduction definitely point to the existence of final causes.

In Aristotle's view, ends are inbuilt and irreducible structural features of reality. In non-living beings, they are specified by nature; in living beings, they are specified by their *psychê*, which is the nature of living beings. *Psychê* exhibits different features in the case of plants, animals and human beings. The ends of the *psychê* in plants limit their operation to nourishment, growth and reproduction. In animals, such ends, while incorporating the intrinsic ends of plants, extend their operation to self-movement and perception. In human beings, while incorporating the capacities of plants and animals, ends extend their operation to *noûs*, which is the unique distinguishing feature of human beings in the animal kingdom. As a result, for Aristotle, ends turn out to work differently in different species. Therefore, as the role played by ends in nature is far from uniform, the correct understanding of ends should concentrate on the way in which these shape natural species driving the development of successful biological features.

6.3 The Principle of the Irreducibility of Forms and Ends

At this stage I will posit a principle that addresses the need to regard ends as irreducible features of nature. Based on Aristotle's hylomorphism, this principle precludes any kind of reduction of ends and preserves the necessity of explaining every part of reality within its specific level, rather than by analysis of more fundamental levels:

> (IFE) The irreducibility of forms and ends. The forms and ends of a structural part of reality are not contained in the material or formal composition of lower structures of reality, nor are they explained by analysis of these structures.

IFE articulates the view that ends are irreducible in a strong sense. The principle commits us to the claim that (i) teleological features are genuinely different in nature and function from the material and efficient causes and that (ii) teleological features cannot be explained by the material and efficient causes of lower structural levels. The physical worldview, or what I earlier called the 'analytic' perspective—the one that concerns the discovery of biological mechanisms—, is useless in understanding the causal conjunction of forms and ends as Aristotle sees it. To the extent that teleological features cannot be adequately analysed in terms of efficient mechanisms, my understanding of irreducibility is strong. It rejects the possibility of explaining teleology as a supervenient feature in the traditional account of supervenience (Davidson 1980, p. 214). This account, which stands out as one possible interpretation of Aristotle's teleology (Charlton 1970, 1985; Broadie 1982; Cohen 1989; Cooper 1982, 1987; Gotthelf 2012) endorses (i) and (ii). Still, the view has not always been presented in the same way. The way in which IFE conceives irreducibility rests on the fact that the higher structures of a living organism, i.e. those visible at the macroscopic level, cannot be explained in terms of or reduced to more fundamental elements despite the fact that physics typically focuses on these latter elements and, when they are taken separately, is usually able to analyse them successfully. Hence, I will suggest that an understanding of ends requires a significant change of perspective as well as a change of crucial ontological commitments.

Just as IFE can be argued for by an appeal to the genuine explanatory framework of higher-level real structures, the principle can equally well be articulated by pointing to the intrinsic limits of what other authors call the 'element-potentials', that is, the potential of the basic elements of Aristotle's natural world (earth, water, air and fire) to embody fully-fledged living beings. In Gotthelf's view, the inherent limits of the element-potentials to constitute living substances justifies the principle of 'ontological irreducibility', which he describes in these terms:

> (…) the development, structure and functioning of living organisms cannot be wholly explained by—because it is not wholly due to—the simple natures and potentials of the elements which constitute these organisms. No sum of actualisations of what I have called 'element-potentials' is sufficient by itself for the production of those complex living structures and functionings for which Aristotle offers teleological explanations (2012, p. 30).

Gotthelf points out that biological explanations are sanctioned by the absence of a full material-level account of the phenomena under consideration in such explanations. In his view, we cannot achieve a full explanation of these phenomena unless we change our perspective in order to unearth an 'irreducible potential for form (…) that provides a primitive directedness upon an end' (2012, p. 71). Everything that can be said of a living organism by an analysis of its parts and its nested relations cannot provide a satisfactory answer about the structure and functioning of the organism. In his view, a full understanding of a living being demands access to a deeper and more fundamental view. The 'irreducible potential of form', then, is supposed to provide this view through a change of perspective. This perspective exposes the inability of a material or efficient analysis to explain the structure and functioning of a living organism as a whole. This explanation requires consideration of its formal features. These features embody a potential source of explanation that remains otherwise invisible to the analytic standpoint, that is, to the analysis of biological phenomena through causally-efficient mechanisms of the kind that are common in biochemistry or molecular biology.

In its support of his view, Gotthelf points to key passages of Aristotle's works (*GA* II 2–3, *Phys.* II 8; *PA* I 1). To quote one of the clearest passages that buttresses his strong irreducibility and the intrinsic limits of the element-potentials, consider this:

> Some think that it is the nature of the fire which is the cause quite simply of nourishment and growth; for it appears that it alone of bodies is nourished and grows. For this reason one might suppose that in both plants and animals it is this which does the work. It is in a way a contributory cause, but not the cause simply; rather it is the soul which is this. For the growth of fire is unlimited while there is something to be burnt, but in all things which are naturally constituted there is a limit and a *logos* both for size and for growth; and these belong to soul, but not to fire, and to *logos* rather than to matter (*De An.* II 4, 416 a 9–19).

Aristotle argues here that activities such as growth and nourishment cannot be simply attributed to the underlying natural elements of a life form. Fire is one of the constituents of bodies and one of the underlying elements of life forms, what he calls a 'contributory cause'. Aristotle notes that, by nature, elements like fire tend to grow unconstrained in every direction unless there is some *logos* that can tame them. The role of *logos* is to make things grow and expand in an orderly and harmonious way, rather than in an irrepressible way that will destroy the organism. Aristotle contrasts this with the lack of *logos* in fire. He notes that the change in the behaviour of the element-potentials inside living organisms underpins capacities like growth. The contrast between the activity and effects of fire inside and outside a living organism clearly indicates that growth and nourishment, as well as most biological capacities, are governed by formal and final causes rather than by material and efficient ones. In Aristotle's physics, the beneficial role of fire in the growth of organisms shows that a counteracting force is powerful enough to effectively tame deleterious side effects.

If, unaware of this, we fail to notice the action of formal and final causes as limits *(logoi)* of underlying natural tendencies, what usually happens is that teleology is unwittingly dissolved into causally-efficient explanations or simply omitted. But in so doing, causally-efficient explanations will assume teleology to make sense of the biological strategies of organisms to rein in the tendencies of the element-potentials to the benefit of the whole. This assumption is not surprising. As argued above, by the time we look at any living organism, the meaningful explanation of the phenomena that take place within it has already settled the fundamental question of whether a substance is living or non-living. When we assume that substance X is alive on the basis of our best evidence, we assume that the properties of certain molecules that combine with some other molecules must ultimately be orientated to growth, adaptation and survival. In light of the assumption that substance X is alive, which must be later confirmed by data, scientific analysis then proceeds to look at its behaviour.

Gotthelf makes a crucial distinction between the element-potentials and the potential for form in *De An.* II 4. In this passage Aristotle shows that explanations that appeal to one or the other are both compatible and necessary, while a one-sided explanation of living phenomena distorts the scientist's outlook and prevents understanding.

IFE is compatible with an explanation of the structure and development of living substances based on Gotthelf's potential for form. But in any case, IFE does not need an analysis of the element-potentials or of their behaviour to be true. This analysis is a scientific task. IFE simply articulates the view that forms and ends are ultimately irreducible to lower structures of reality because forms and ends are real dimensions that require independent explanation. This explanation will not be satisfactory unless we change from the perspective that enables the discovery of causally-efficient mechanisms to the perspective that enables the discovery of teleology. If we do not abandon the causally-efficient perspective, at least momentarily, we will fail to understand the reason why any material organisation from eukaryotes to mammals is intrinsically more than a happenstance ensemble of molecules whose interactions give rise to capacities like proportionate growth, adaptation and survival. In other words, we will ultimately fail to understand the hidden or underlying assumptions of most scientific theories.

To see how IFE can change our understanding of science, let us go back to the example of the mosquito. We can contemplate the mosquito's desire for food as an end. A sophisticated sensory apparatus enables mosquitos to do what animals and some birds cannot easily do: to accurately locate sources of heat in the dark. Once we know that the mosquito can detect sources of heat, we see the reason *(logos)* of its nocturnal foray. Given that the mosquito has active sensory organs, its activity becomes relevant to our account. Once we find out that the mosquito *detects* sources of heat, it is reasonable to think that this detection should feature as part of a complete explanation of its nocturnal foray, for such detection cannot be illusory. The fact that mosquitos detect heat is a crucial underlying assumption for the biological analysis of a mosquito's organs. If the detection of heat is part of the way in which the mosquito perceives, we may infer that the mosquito should have organs apt for its needs. In other words, we may assume that the need of food should guide the analysis of the perceptual capacities of the mosquito. On this basis, biology examines the mosquito's perceptual apparatus in search of clues of how it interacts with the environment. Accordingly, knowledge of the infrared capacities of mosquitos is embedded in cognitive assumptions about both the perceptual abilities of the mosquitos and the link between their perceptual abilities and their basic needs, all of which assume that the mosquito can succeed in satisfying its needs.

If the detection of heat were a mere heuristic principle *a la* Kant, that is, a guiding principle posited merely to generate working hypotheses, a plausible explanation but without any ontological commitment to existing natural ends, the discovery of the mosquito's sensory apparatus would be merely coincidental. Only when the connection between the detection of heat and the enabling organs of the mosquito is real rather than hypothetical or heuristic can the scientific analysis of its organs be accurately guided towards those organs that explain its nocturnal foray as opposed to those organs by which the mosquito tries to stay awake. The scientist knows that the connection between the attraction of heat and the enabling organs of the mosquito is not merely coincidental or contingent. In this context, while the mosquito's ability to detect heat is crucial, science has no good explanation of it for two reasons. First, sensitivity to heat cannot be said to act on the design of infrared capacities. Rather, scientific explanations usually work the other way round: assuming that the mosquito has infrared capacities, it is usually suggested that it can detect sources of heat and direct its flight towards them. Secondly, while perception in invertebrates such as the mosquitos might not justify our attribution of desires and feelings to invertebrates, this attribution seems better justified in mammals. Therefore, the lack of a basis for the attribution of perception to some invertebrate species does not seem a valid argument against the analysis of perception as ends.

There are other fundamental reasons why, despite playing no visible or otherwise apparent role in scientific accounts, there is a case for the existence of ends. Nagel famously argued that desires and feelings cannot be described in the objective terms of science and do not play any role in them. This impossibility does not entail the inexistence of subjective properties. To the extent that science restricts its scope to objective features, it lacks the capacity to capture what it is like for an insect or a bigger animal to perceive. Certainly, it may be questioned whether the perception of mosquitos and its drive for food is really conscious or whether this question may actually have a definite answer. But this question raises a separate discussion that we cannot address here. The point is: here is a portion of the mosquito's ontology that science cannot reckon with. While science can work upon the assumption that the mosquito is sensitive to heat and thus explain its perceptual apparatus in the objective terms of biology, if the perception of the mosquito were to be possibly conscious, it would turn out that biology fails to explain facts about the mosquito that have real ontological significance (Nagel 2012, p. 35).

We can say more about the mosquito's sensitivity by conceiving it as an end that resists reduction by scientific analysis. In this way, the mosquito's perception will require independent explanation. In line with the contents of IFE, we can consider a mosquito's detection of heat as an irreducible end that invites action to satisfy a need of food. Since we can regard this need as an end that demands satisfaction, the actions taken by the mosquito towards its satisfaction are the means to a specific end. This end cannot be broken down to the perceptual organs of the mosquito, even if a causally-efficient analysis of such organs turns out to reveal unprecedented knowledge of how the perceptual apparatus of the mosquito is activated by nearby sources of heat. The teleological perspective sees this attraction embedded in the causally-efficient analysis of its perceptual apparatus. In this way, the detection of food, considered as an end, captures a distinctive fact about the nature of the mosquito. This fact is 'efficient' or causal in Aristotle's sense, because it precipitates an action that repeats itself not just in other forays of the mosquito, but also in many other neighbouring species. However, this efficiency is significant for us because the detection of food works as a trigger. The capacity of the mosquito's organs cannot constitute a complete explanation of its foray, because this capacity does not ultimately explain why the behaviour of the mosquito repeats itself so much. In this context, the assumption that the mosquito is capable of perception is not idle or irrelevant, because it has a crucial role to play to effect the required change of perspective. In Aristotle's view, just as the design of an artefact guides its staged manufacturing in men, in higher animals feelings of desire guide the satisfaction of certain species-specific ends. In our account, the detection of food works in the way in which the feelings of higher animals guide their pursuit of goals.

Since science cannot dispense with mechanistic explanations, IFE claims only that these cannot yield a 'full understanding' of nature, that is, knowledge of the first causes of reality in Aristotle's sense. Examples of the phenomena that remained unexplained for Aristotle until viewed teleologically are the growth of roots and the generation of animals. But as the example of the mosquito has shown, the same point applies to the understanding of life forms, of perception, of feelings, and of other irreducible phenomena that we usually grasp through the comprehensive view of teleology.

To do justice to Aristotle's philosophy of biology, teleological dimensions should be seen as fully compatible with modern science. I can see no conflict between IFE and scientific explanations, because both of them focus on different facts or dimensions of a nature that is one, the same and continuous. Aristotle's philosophy of biology presupposes this distinction of perspectives, even if his explanations often conflate them so that it is hard to tell biological from philosophical claims, since his biological claims often come laden with concepts that clearly transcend biology.

The more implausible hypotheses of Aristotle's physics and biology that reflect the limits of scientific understanding of his time should not be viewed as an obstacle to reconciling his teleological views with modern science. His theories would be quite different today if informed by knowledge of modern science. When Aristotle argues that fire is not the cause of nourishment and growth (*De An.* II 4, 416 a 9) he is not denying a scientific fact. He is rather arguing that Empedocles was wrong in attributing the growth of a plant to the mechanistic action of its constitutive elements; more was clearly needed. Accordingly, if Aristotle had known that growth is a controlled process of cellular division that is unrelated to fire, he would likely have claimed instead that hormones are not solely responsible for growth. In other words, he would have argued that cellular division could not yield a full understanding of the process of growth, since this understanding points to the role of growth in the larger context of a living organism. But in doing so, teleology is no replacement for a scientific analysis of the role of hormones or of cellular division in growth. Its function is rather to point to the principles that regulate growth and give it *logos*, that is, both a sense of proportion and a definite direction. Only if we regard teleology as guiding or orientating scientific discovery, rather than hampering it, will we appreciate the rich variety of biological phenomena and succeed in freeing biology from the clutches of physicalism.

References

Barnes, J. 1985. *The complete works of Aristotle, 2 vols*. Princeton: Princeton University Press.

Broadie, S. 1982. *Nature, change and agency in Aristotle's physics*. Oxford: Oxford University Press.

Charlton, W. 1970. *Aristotle's 'physics' book I and II*. Oxford: Clarendon.

Charlton, W. 1985. Aristotle and thc harmonia theory. In *Aristotle on nature and living things: philosophical and historical essays in honor of David M. Balme's seventieth birthday*, ed. A. Gotthelf. Pittsburgh/Bristol: Bristol Classical Press.

Cohen, S.M. 1989. Aristotle on hot, cold, and teleological explanation. *Ancient Philosophy* 9: 255–70.

Cooper, J.M. 1982. Aristotle on natural teleology. In *Language and logos: studies in Ancient Greek Philosophy presented to G. E. L. Owen*, ed. M. Schofield and M. Nussbaum. Cambridge: Cambridge University Press.

Cooper, J.M. 1987. Hypothetical necessity and natural teleology. In *Philosophical issues in Aristotle's biology*, ed. A. Gotthelf and J.G. Lennox. Cambridge: Cambridge University Press.

Davidson, D. 1980. Mental events. In *Essays on actions and events*. Oxford: Clarendon.

Gotthelf, A. 2012. *Teleology, first principles, and scientific method in Aristotle's biology*. Oxford: Oxford University Press.

Johnson, M.R. 2005. *Aristotle on teleology*. Oxford: Oxford University Press.

Kim, J. 1993. *Supervenience and mind: selected philosophical essays*. Cambridge: Cambridge University Press.

Leunissen, M. 2010. *Explanation and teleology in Aristotle's science of nature*. Cambridge: Cambridge University Press.

Lewis, D. 1966. An argument of the identity theory. *Journal of Philosophy* 63: 17–25.

Nagel, T. 2012. *Mind & cosmos. Why the materialist Neo-darwinian account of nature is almost certainly false*. Oxford: Oxford University Press.

Papineau, D. 2002. *Thinking about consciousness*. Oxford: Oxford University Press.

Papineau, D. 2009. The causal closure of the physical and naturalism. In *The Oxford handbook of philosophy of mind*, ed. B.P. McLaughlin et al. Oxford: Oxford University Press.

Stoljar, D. 2009. Physicalism. In *The Stanford encyclopedia of philosophy*, ed. E.N. Zalta. http://plato.stanford.edu/archives/fall2009/entries/physicalism/.

Chapter 7
Body, Time and Subject

José Ignacio Murillo

7.1 Introduction

"I am my body." At first glance, this statement might seem obvious. It is not so easy, however, to clarify its meaning. Doing so clearly obliges us to explain what it is that we understand by "body," and also presents us with the task of defining the identity that this sentence expresses between "I" and the body which I can call "mine."

The above statement seems incompatible with another that is also common in our language: "I have a body." In this case, a rift is opened up between the "I" – the self – and the body; a distance between them is admitted, which a strange relationship of possession attempts to overcome. As Wittgenstein pointed out, anyone who utters this sentence can be asked: "Who is speaking here with this mouth?" (Wittgenstein, *On Certainty*, §244. Quoted by Carman 1999).

That said, if we stop to consider the first sentence again – "I am my body" – we realize that the relationship of possession is also implied here, in the possessive determiner "my," which introduces the distance that one is attempting to surmount in identifying the "I" with the "body." Without presupposing such a distance, making reference to possession – even in this weak form – would be meaningless, at most a simple metaphor.

Are we obliged, then, to accept that neither of the two expressions makes sense? One could argue that this confusion, which our language encourages, is at the root of the problems that lead to dualism. It's not as simple as that, however. If the identity between me and my body must be affirmed, it is because it is not, at a surface level, evident. "The distinction between subject and object is blurred in my body" (Merleau-Ponty 1964, 167; quoted in Carman 1999, p. 206). But it is also true that

J.I. Murillo (✉)
Mind-Brain Group, Institute for Culture and Society (ICS), University of Navarra, 31009 Pamplona, Spain
e-mail: jimurillo@unav.es

M. García-Valdecasas et al. (eds.), *Biology and Subjectivity*, Historical-Analytical Studies on Nature, Mind and Action 2, DOI 10.1007/978-3-319-30502-8_7

I experience what I call "my body" in such a way that cannot totally substitute for my subjectivity. Furthermore, if we declare the distinction meaningless, do we have to accept that scientific descriptions about my body will never be able to have a direct relationship with the way I see myself, in my consciousness or my subjectivity? If this is the case, and it turns out that I am not my body, what am I?

To return to the first expression: "I am my body." What do we mean when we make this kind of identification? Are we trying to eliminate the "I" by burying it in the corporeal? Or do we rather aspire to make our bodies "I"s, to turn them into subjects? The distinction between the "top down" and "bottom up" approach – as they are sometimes referred to – is relevant at this juncture. I propose to spend some time examining this difference, given that the latter approach allows us to better test the claims of naturalistic and reductionist biology.

The sentence: "I am my body" seems to be a corollary of the thesis that gives biology the last word when it comes to describing and explaining what a human being is. But note that the approach I have adopted until now does not exactly coincide with the "mind-body problem," as it is usually understood: the relation between biological accounts of our organism and subjective experience. For the latter the body is no longer seen from the point of view of my direct experience, but as a scientific object, and the same happens with subjectivity. That presupposes, in turn, a theory about the brain and, to some extent, about its relation to subjective experience considered as a type of object.

However, with both views we see that interiority and exteriority are always intertwined when we speak about the human being and the human body. The approach I have chosen as a basis for these reflections tries precisely to warn against the danger of a hasty exteriorization and objectification of the terms of the problem. Nonetheless, in both views we see that interiority and exteriority are two perspectives that are difficult to avoid when we speak about human beings.

7.2 Internal and External Perspectives on the Body

Phenomenology has particularly emphasized the difference between two ways of considering what is referred to in English as the "body" (Welton 1999). On the one hand, we may consider our bodies to be one of the many bodies that populate the universe, and attempt to understand them regardless of the fact that they are uniquely our own, using a description that we can call "objective" or "external." This is the lens through which the body is usually studied by the natural sciences. On the other hand, we can also understand the body from an internal perspective as a "lived body," the German *Leib*, which some translators render as *soma* and others as "personal," "sentient" or "sensitive body" (Behnke 2004, pp. 238–239). If we adopt this position, bodies are presented unequivocally as our own, as extensions or realizations of our selves, which nevertheless should not be totally identified with them.

Which of these two perspectives is primary? If it is a case of answering the question of how I come to conceive a body as my own, it seems that the second would have to be affirmed as being more fundamental. How could we otherwise know what such a thing might mean, however much we inspected living or lifeless bodies from the outside?[1]

There do exist ways of attempting to explain the internal experience of a body from an exterior and objective point of view. For example, we might attribute the peculiar double aspect or double point of view of bodies to the existence of various types of sensory receptors. The external receptors, in this view, are those which obtain information about the environment. For them the body is transparent, it does not appear directly, because they aim to fulfill a different function. In addition, we find internal receptors, which respond to a set of nerve fibers responsible for providing us with information about our body (Craig 2003). If we ask ourselves what is the foundation of this distinction between the internal and the external, between what is within and without, we may conclude that it is necessary for the nervous system to distinguish self and other in order to serve the purposes of the organism. What is "within" is what is necessary to protect, and nothing more. Precisely because the organism is a body, its existence depends on its ability to self-identify and distinguish between what composes it and everything else. The boundary between myself and the other would therefore be a consequence of the demands of survival.[2]

On the other hand, the thesis that biology should establish the framework for anthropology and the study of living beings in general has been criticized by Heidegger, among others. In *Being and Time*, Heidegger argues that we obtain our concept of life through a de-potentiation of Dasein. The opposite – attempting to understand life by examining how it is manifested in other beings – is an impossible task (Heidegger §10). Consequently, according to Heidegger, internal experience is crucial not only in order to understand ourselves, but also to understand the rest of reality.

Hans Jonas, who was a doctoral student of Heidegger, made a similar posture into the center of his philosophical project. He suggests that our internal experience is the basis for our comprehension of life in general, the main analogue of our knowledge of living beings. This inner experience acquires central gnoseological relevance in Jonas's work, as it provides the only way we can reach an understanding of what it means to have inner being in general. It makes the meaning of such notions as life, substance and even causality accessible to us (Jonas 1992, 2001). Only through the experience of "being me," as it is presented in inner experience, can I understand other living beings as such, identifying them in their particular and irreducible identity (*Selbheit*, *Selbigkeit*). Understanding life seems to require a kind of empathy, the capacity to partake in the same laborious struggle to continue

---

[1] From a phenomenological perspective, the importance of corporeality as an inherent determinant of all experience has been highlighted in a radical – and, in my view, somewhat one-sided – way, by Hermann Schmitz (2011). He goes so far in his attempt to avoid the dualism and mentalism of much of the Western tradition that the distinction we have been discussing between the interior and exterior perspectives seems to vanish.

[2] A philosophical example of this view is the *Self-Model Theory* of Thomas Metzinger (2003). For a critical survey of this theory, see Murillo (2011).

living and free oneself from the servitude to matter which apparently threatens life at all levels. In addition, the knowledge of oneself as the cause of certain actions, together with the experience of the resistance with which our intentions are met in reality, opens up the possibility of using causality as a basis for understanding the relationships between beings in the universe.

In defending this thesis, Jonas might seem to be an heir to the European idealist tradition that includes figures such as Schelling (2000). In idealist philosophy, self-consciousness is the starting point of philosophy, and nature is rational in so far as it constitutes a "sketch" or "precedent" of subjectivity. However, Jonas does not intend to subscribe to any form of subjectivism or idealism, but rather proposes a particular kind of naturalism. Our perception of life gives us a privileged vantage point from which to understand the universe, which, as manifested in us, is active, intelligent and free. Our experience thus precludes any attempt to understand the universe by excluding the "phenomenon of life," which for human beings is so obvious and unavoidable. As Jonas argues so eloquently, the relevance of this phenomenon is both epistemological and ontological. If the universe allows life to arise within it, any understanding we have of the former must include the possibility that life exists with all its attributes. The internal experience of human beings and their particular way of living thus become the basis for understanding the universe and an indispensable reference point for judging any metaphysical conception of reality.

These attempts to escape dualism show that interiority and exteriority are two aspects that are hard to avoid in dealing with living beings, and with human beings in particular. Do we have to decide between these two points of view? Or must we instead accept that neither of them is totally reducible to the other? In the latter case, what consequences follow for our understanding of the human person as a living being? Is there any point of view that can present these two aspects as at least being compatible, or even as mutually enriching?

7.3 The Dual Aspectivity of Living Beings

Helmuth Plessner, meanwhile, goes down a different path in his explanation of the emergence of this duality. He does not take internal experience as a starting point for a philosophical approach to living beings. His intention is to grasp the distinctive character of life on the basis of the way in which it appears to us. And he thinks that it is precisely this within-without duality, which is present in all forms of life, that distinguishes living from non-living things.

What is especially valuable in Plessner's proposal is that he maintains this duality while avoiding both the risk of substantializing the poles of internality and externality and the risk of reducing either one to the other. To understand life, he suggests that rather than focus these poles in themselves we should examine the boundary itself. The distinctive feature of life is, in fact, the particular relationship that living beings establish with their surroundings by means of this boundary (1981).[3] While for lifeless beings we understand a boundary as the separation between two bodies, that is, the neutral place where one thing ceases to be itself and something else begins, in the case of living beings the boundary belongs to the living being itself. It marks the distinction with respect to its environment. For Plessner, this "dual aspectivity" (*Doppelaspektivität*) is the hallmark of living beings. The living being, according to him, is not simply a *gestaltic* whole, in which the perception of the whole is prior to or simultaneous with the perception of its parts, but rather something more: living beings actively distinguish themselves from an environment that they cannot live separately from, and take the initiative in the exchanges that they establish with it.

Using this boundary as a starting point, Plessner posits three modalities of the organization of living existence, taking the relationship they establish with their boundary as the criterion. The first is that of the plant, which he calls *open* (1981, p. 282f). In animals, on the other hand, we find a *closed* form of living existence, characterized by *centrality* and *frontality* (1981, p. 291). In an animal, the body is, as it were, asserted against the environment, which in turn acquires a new meaning. The animal both is its body and is in its body (1981, p. 303). This centrality increases in some life-forms, and in the human being takes a new turn.

In contrast to the centrality present in animals, the human being is characterized by *ex-centricity* (1981, p. 360f). As in the cases of plants and animals, this is consistent with the configuration of our body. Human existence is dual because it develops at the boundary. When we internalize this dual aspectivity, appropriating the boundary in both directions, the environment becomes the *world*. The human being not only perceives what surrounds her from her position, and values it according to the reference system imposed on her by her vital interests. In addition, she is also capable of perceiving herself from the outside, from an external perspective. Plessner warns us that, from the fact that the human being passes from living inside her corporality to living outside it, an unavoidable duplicity and a real fracture in her existence arises. Her unity is only a mediation. In this way, according to Plessner, the human being has a triple determination: the person is her body, but is also in a body (as interior life or soul) and, at the same time, outside the body.[4]

It is interesting to note how Plessner's observations fit with aspects of our bodies that have only recently come to light. When it comes to identifying that organ that plays the greatest role in determining our individuality, and which is the most characteristic of the human species, we usually point to the brain. The fact that this organ is found hidden inside the skull is associated with the idea that the mind is the

[3] "Körperliche Dinge der Anschauung, an welchen eine prinzipell divergente Außen-Innenbeziehung als zu ihrem Sein gehörig geständlich auftritt, heißen lebendig" (1981, p. 138).

[4] "Positional liegt ein Dreifaches vor: des Lebendige ist Körper, im Körper (als Innenleben oder Seele) und außer der Körper al Blickpunkt, von dem aus es beides ist. Ein Individuum, welches positional derart dreifach charakterisiert ist, heißt *Person*. Er ist das Subjekt seines Erlebens, seines Aktionen, seiner Initiative. Es weiß und es will. Seine Existenz ist Wahrhaft auf Nichts gestellt" (1981, p. 365).

seat of internality, a sort of CEO that communicates with the exterior world through the rest of the body, using it as a tool. However, during embryological development, the brain originates from the same group of cells that produce the skin. This suggests another possibility: understanding the brain as a mediator between the external and the internal.[5]

The embryological relationship between the brain and the skin allows us to understand the brain in the context of the mediation between the living being and its environment, and makes clear the importance that the separation-communication duo has for the human being, understanding separation here as distinction rather than isolation. The skin is not just a protective casing, but rather forms part of the system of distinction and initiative that characterizes the organism. In the case of humans, this boundary is a place for experience and learning via the senses. The brain is the organ most sensitive to these exchanges, as it constantly reconfigures itself in view of them. At the same time, the brain is responsible for changes in behavior, which is the specific way in which animals interact with their environment.

Plessner represents an attempt to reconcile the internal with the external perspective without allowing them to dissolve into each other. From the platform of his external description of living beings, he manages – relying on dual aspectivity – to open up a space for the inescapable inner experience which imposes itself on us as an essential feature of life, even in its simplest forms.

Whatever one make think of Plessner's position in its entirety, he makes a convincing case for the inseparable duality of the two perspectives regarding our bodies. In reality, although the difference between these perspectives is evident at first glance, it is not at all clear that they are completely independent. A full experience of the body as one's own does not appear to be possible without a certain knowledge of it from without. It is in this way that we fully grasp the consistency, strength and limits of the body, and how it appears before others. Otherwise the body could not be properly used by the living being as an instrument or means of expression. Furthermore, the knowledge I have or the ideas I develop about my body and other bodies are intertwined with the way in which I perceive my body to be mine, and both are certainly influenced by psychological and cultural factors.[6]

[5] Fuchs calls the brain a "mediating organ" (2013).

[6] Bruno Snell (2000, pp. 13–20) points out that in ancient Greek art and literature the body is not seen as an organic unity until the fifth century BCE. Previously the body was seen as different parts added to each other. They were *melea kai gyia*, that is, parts with muscle, strong, articulated and able to move. The word *soma*, however, that corresponds to our term "body" was only used by Homer as meaning "corpse." What we call "body" is referred to in plural as a collection of members and is considered from the point of view of human activity.

7.4 The Living Being and Temporality

The identification of the self with the body faces a further problem. The experience of the self does not merely suppose that I identify myself in some way with my body, but also that I do so over time. From both an external and internal point of view, the body is subject to change. For external experience, the permanence of the self must be identified with the persistence over time of the body which is related to it. This poses the problem of the identity of a living being as it undergoes change.

From the scientific point of view we can try to grasp the complex organization that is characteristic of living beings as a special kind of *system* and the changes and evolution they experience as kinds of *processes*. The notion of "system" expresses a unity arising out of what is complex, a unity which can remain present throughout change. When we call something a "system," we assume the existence of an interrelated diversity that can be formalized.[7] The science that formalizes such relationships, however, is mathematics.[8] This conclusion makes us see that understanding the living being as a system, rather than understanding on its own terms, steers us toward a mathematical approach.

The notion of "system" is close to that of "law," the quest for which is the distinctive goal of modern science (Murillo 1999). There are many types of laws in science, but all of them describe relationships between different aspects reality —most often, if possible, measurable properties. Although we can identify a systematic unity in a whole thing that moves, that unity is always a stable formality or relationship made up of various elements. So the notion of system is akin to that of "number" and "function." In fact, a number is a formal relation between elements that is independent of the nature and characteristics of the elements. And a mathematical function is in this way a "relation of relations," because it establishes a relation between numbers independently of which numbers are involved.

A system can persist even though its elements or parts vary or are substituted, provided they fulfill their role in the mutual relationships that defines the system. To say that the unity of the living being is that of a system which persists over time, however, raises the problem of identifying the system. It is one thing to accept the common perception that something persists, and quite another to say that this or that particular part or aspect is what does not change. To this we may also add the problem of identifying the living being and its unity with a kind of relationship – the system – which, as with all mathematical objects, is in itself inactive.

One way to include time in living beings is to consider the living being as united to its whole life cycle. From this perspective, a living being has temporal parts which may not be totally present at any given moment, because what we identify as

[7] "A system can be defined as a complex of interacting elements" (von Bertalanffy 1969, p. 55).

[8] This affinity between the notion of system and mathematics is especially evident in the cited work of von Bertalanffy (1969).

something "here and now" is only an instant in an ongoing temporal process. From this perspective, change is but variation *in* time.[9]

However, identifying the living being with a process or its temporally extended life processes also presents difficulties. First, it is not consistent with our natural attitude towards living beings, which accepts that they persist over time; and as a result I can identify one of them now as something real, independently of the alterations it may experience in the future. In other words, it seems that living beings exist insofar as they create and preserve their own "now," at least to some extent. Moreover, this approach poses serious problems for the identification of a living body with a self. I do not say of myself that I am a process, but rather that I am *now*. For example, I am the one who is acting at this moment, independently from any possible future. If it were not possible to affirm this, the idea of the self would be useless.

This approach also assumes that time can be considered in the same way as space. But the spatialization of time, while useful for calculation, does not grasp time as it really is. It gives the impression that time is some kind of preexistent container wherein things are placed, with inclusion of their "temporal parts." But on the other hand, if we accept that a being is real in each instant, we face the question of how it relates causally to the existence of itself at a future time.[10]

Turning to internal experience, we also face problems when we try to identify the self with the body over time. The body that I perceive to be mine is changing, but the sensory perception that sustains this experience is also changing. We can identify patterns that allow us to connect perspectives that follow each other in time; for instance, in this way we might recognize a hand by the way some of its features belong to some kind of perceived pattern. But it is not so easy to explain how we can say that the perceiver is the same as whoever perceived beforehand. Sensory patterns – what Aristotelian psychology would call "images" – are always applied to sensory content qua something external. Knowledge of the continuity of the self, however, is not simply recognition; even less is it equivalent to the recognition of this body as my own over time.

Moreover, the identification of this body as mine does not exhaust the reality of the self – quite the contrary. Rather, if we disregard the process by which the self develops and comes to being and we focus on the immediate experience of the self in activities like knowing or willing, the "I" could appear independently from the body, in the case of knowing, and in the case of willing, its existence is assumed only afterwards, as a condition of execution. Given that my experience of subjectivity is essentially linked to these experiences, selfhood seems to depend not so much

[9] "Change is analogous to spatial variation. Change does occur in virtue of unchanging facts about temporal parts." Sider (2001, p. 214).

[10] "The organizational account of functions relies, then, on the theoretical assumption according to which the various temporal instances of a system, in spite of any changes which may occur, can be considered as instances of the same encompassing self-maintaining organization, to the extent that their constitutive organizational properties are causally transmitted from one instance to another instance by the maintenance of a material connection between them" (Saborido et al. 2011, p. 599).

on my body as on the exercise of activities in which the role the body plays in their execution can only be grasped secondarily.[11]

7.5 Temporality and Subjectivity

It is easy to understand why Kant proposed time to be the a priori form of internal sensory experience.[12] Time itself seems to allude to subjectivity and consciousness. Hegel noted that the opposition between space and time is the first and most basic opposition to appear in the philosophy of nature. Space for Hegel is the most extreme alienation from the idea, a "bad infinity" into which the mind cannot delve. On the contrary, time at this level proves to be the first appearance of subjectivity (Hegel 1992, §42). And it is subjectivity which guides the process by which nature becomes spirit, which is for Hegel the only point of view from which nature is rational, i.e. intelligible. This Kantian concept of time is the empty time of sensibility, a mere flowing. For Hegel, however, the most important dimension of time is the present. From this perspective, time is more subjective than space – the very image of exteriority – because within time there appears, albeit in a lightly sketched, provisional way – the image of the uniformity and constancy of pure "passing" – the present.

As is well known, Heidegger took up the Kantian subject of time with great originality, presenting a new understanding of time, and courting a clash with Kant as much as with Hegel. Heidegger objects to Hegel on the grounds that his vision of time ends up destroying it, and proposes a different view of time. This he bases on a human trait that is, for him, a fundamental ontological phenomenon: care, or concern (*Sorge*) (Heidegger 1996, §42). From this point, time is seen as an "ecstasy" in which the present is but one dimension, and not the most important. For Heidegger, the critical dimension of time, as the analysis of Dasein reveals, is the future. This is, in his opinion, a crucial feature of his critique of modern subjectivity.

Two topics appear clearly in this discussion: on the one hand, the connection between presence and subjectivity, and, on the other, the connection between time

[11] In a similar way Scheler distinguishes between functions and actions: "First, all functions are ego-functions; they never belong to the sphere of the person. Functions are psychic; acts are non-psychic. Acts [loving, hating, judging, for example] are executed; functions happen by themselves. Functions necessarily require a lived body and an environment to which the 'appearances' of functions belong. But with the person and acts we do not posit a lived body; and to the person there corresponds a world, not an environment. Acts spring from the person into time; functions are facts in phenomenal time and can be measured indirectly by coordinating their phenomenal time-relations with measurable lengths of time appearances given in functions themselves. For example, seeing, hearing, tasting, and smelling belong to functions, as do all kinds of noticing, noting and taking notice of (and not only so-called sensible attention to) vital feeling, etc. (…)" (Scheler 1973, p. 388). Concerning the different ways of experiencing the self in knowledge and action, see Murillo (2009).

[12] "Time is nothing other than the form of inner sense, i.e., of the intuition of our self and our inner state" (Kant 1998, A 33, B 49).

and presence. Presence is, in fact, an essential feature of what we call consciousness, whatever its content may be. However, we cannot avoid the preeminence of subjectivity by dissolving presence in time, because time also refers to subjectivity.

Plessner criticizes Heidegger, arguing that his point of view is still affected by the modern concept of subjectivity, which sees the human being as the closest to himself. Plessner's proposal, however, rests on the conviction that the human being is distinguished from all other beings in that he is neither the closest nor the farthest from himself (1981, p. 12). From this point of view, the centrality of temporality may also be considered as the consequence of granting privilege to the internal perspective, that of subjectivity, when it comes to understanding life.

Despite his profound revision of modern subjectivity, Heidegger falls victim to this critique. In the end, from Heidegger's position it seems that the only valid, and even possible, perspective is the internal. Obviously, one of the negative consequences of the position that Plessner criticizes is the difficulty of integrating the body as it is studied in the sciences – biology in particular – into the philosophical study of man.

At this juncture it may be helpful to consider what the Aristotelian perspective has to say about time. For Aristotle, time is not a primary notion that can be understood by itself. Time is "the measurement of movement according to a before and an after." (Aristotle 1983, IV, 220 a 24–25). It may be objected that this definition implicitly includes time and is thus circular. However, the observation that time refers to something distinct from itself is worth considering. There can be no time without movement. Moreover, Aristotle's definition also indicates that movement itself is not enough to speak of "time." A measurement is necessary, and this requires a comparison. It is the soul which introduces the "now" and is the reference point from which the before and the after emerge. In other words, we can say that Aristotle agrees with Hegel that it is the mind that establishes the present. But unlike Hegel, for Aristotle this present is not destined to nullify or abolish the other dimensions of time. In other words, movement, understood from the mind's perspective in the present, is the real distinction (not merely ideal or thought) between the before and after.

7.6 Movement, Operation and Time in an Aristotelian Approach

Does Aristotelian psychology have anything to say about the experience of the body? More specifically, does it take a position on the priority of the internal or the external perspective for understanding the living body?[13]

[13] In addition to following the Aristotelian text here, I will also integrate Leonardo Polo's proposal, which declares itself to be a continuation and further development of Aristotelian philosophy. See Polo (2003, 2006).

It is true that Aristotle does not consider the problem of internal and external perspectives in the same way that contemporary philosophy does. The familiar form of this problem derives from the sheer opposition that we experience between the objective way in which empirical science deals with reality and the subjectivity that is left over (and which questions the authority of empirical science as the only rational method for the study of nature, human beings included). But this does not mean that Aristotle does not have anything to contribute to this discussion. As we will see, his contribution concerns precisely that which Jonas and Plessner also hold to be key: the concept and reality of life.

At least two characteristics of the Aristotelian approach to life are especially relevant for our purposes. First, this approach focuses on vital activities rather than living bodies. Indeed, Aristotle is even willing to accept the possibility of living beings that are not bodies. The second characteristic, which is intimately connected to the first, is an implied theory of movement and time.

Aristotle, in an important and often overlooked passage of his *Metaphysics*, distinguishes two kinds of movements (Aristotle 1924, IX, 6, 1048 b 18–36). Here he says that movement cannot be defined and properly understood without reference to a *télos* or end. This holds for more than just consciously oriented, purposeful processes: from an external point of view as well, movement cannot be defined without reference to a goal. This is true for all kinds of processes, even for local movements. They all are oriented, coming from somewhere and going towards somewhere. All movement cannot be properly understood if we do not consider it from the point of view of the natural inclinations of the thing that is moved.

To speak about such natural inclinations, especially in the case of non-living beings, may seem like a regression to a conception of reality that has been totally discarded by modern science (Spaemann and Löw 2005). However, in my opinion this critique is based on the confusion of the point of view of modern science, as exemplified in Newtonian mechanics, and the consideration of natural beings as they are in themselves. The latter is precisely the point of view of classical natural philosophy. Every natural being has a *nature*, which is, as Aristotle says, the principle of movement and rest within each thing (Aristotle 1983, 1, 192b20–23). Natural beings are not only characterized by their properties; they also tend to move in a certain way. This internal principle of movement is the key to distinguishing between natural and violent movements. Something is moved in a violent way when it is lead away from the movement that internally and spontaneously arises from it in a certain context.

So reference to a *télos* or end is essential to understanding movement, especially when we consider it as an internal affection or activity of natural beings, and not from an exterior and formal point of view. And it is precisely from this point of view, from the internal relation of each movement to its *télos*, that Aristotle establishes his most relevant contribution to the understanding of living beings: the distinction between processes (*kínesis*) and activities (*enérgeia*). *Kínesis* is the kind of movement whose *télos* is external. Aristotle gives the construction of a building as a clear example. Construction is a process oriented to a result, which is at the same time the *télos* (of the builder, in this case) and the limit (*péras*) of the action. The builder cannot build once the edifice is finished. The activity of building is limited. It is oriented towards something external to itself and cannot persist when it comes to its conclusion. Its *télos* does not belong to it. But this is also the case of other processes, even of local changes: movement as activity ceases when something arrives at a resting place.

On the other hand, there are some "movements" which are called "operations" or "perfect activities" (*praxis teleia*) because their end is internal to them. They in fact consist in the possession of their ends. And this is the case, for example, of seeing. I see and, at the same time, I have seen, so I can continue seeing. There is no internal limitation to the activity because the end to which it is oriented is not outside the activity: it is reached in the same measure that the activity is executed. Living and all vital movements are, for Aristotle, activities (*enérgeiai*), because they are possessive movements, as is life in itself: living is not in itself a process with a result, but an activity which has in itself its own objective.

This approach to life can now be compared with Jonas's and Plessner's position. An Aristotelian characterization of interiority would be developed in terms of perfect and possessive activities. Only a being which can reach and possess a goal in some way can establish a real distinction between interiority and exteriority.[14] We cannot understand life strictly from the point of view of its products or realizations. The distinctive trait of life is its capacity for possessing the good (*télos*) and for improving (Murillo 2008). As a result, the boundary between the corporeal living being and its environment can be seen from a more positive perspective, that is, not as an unstable mediation but as derived from a kind of activity which is more fundamental than the boundary.

7.7 An Aristotelian Perspective on the Experience of the Body

The simultaneity of activity and goal in vital activities is not realized by excluding movement but by giving it a new orientation – the interiority of living beings – and opening the possibility of a synchronization of movements, which is the kind of unity uniquely enjoyed by organic living beings.

A particular case of synchrony appears in sensitive knowledge, which is characteristic of animals and human beings. Our senses receive stimuli from the outside world and respond with a cognitive activity that possesses reality through intentional objects such as colors, sounds, etc. Although external sensory experience

[14] Medieval Aristotelianism usually translated this distinction between perfect activity and movement as *inmanens* and *transitiva*: "Action is twofold. Actions of one kind pass out to external matter [*transiens*], as to heat or to cut; whilst actions of the other kind remain in the agent [*inmanens*], as to understand, to sense and to will. The difference between them is this, that the former action is the perfection not of the agent that moves, but of the thing moved; whereas the latter action is the perfection of the agent" (*Summa Theologiae*, I, q. 18, a. 3 ad 1).

cannot happen without the body, in such experience the body tends to go unperceived; it becomes a mere vehicle of knowledge. The body determines what I can feel, as it makes perception possible. The changes in its internal states and its place and position in space makes sense experience changing and unstable. The only way that sensory experience may be useful for guiding behavior is when it is integrated and stable. In classical psychology, this is the role attributed to the internal senses: common sense, imagination, memory and cogitation.

Common sense is the capacity to feel all kinds of sensory activity, and makes it possible for the sentient being to integrate all the objects of external sensitivity as sensations that can be compared with one another on a common ground. This activity makes the act of imagination possible: it makes patterns and regularities objective, through which it is then possible to interpret and recognize what one perceives and provide an adequate response. None of these acts, however, is responsible for the subjective condition of experience. The subject is not manifest in these acts as such.

To these faculties, which we could call purely objective, we must also add the cogitative sense[15] – the sensory ability to grasp and judge the significance of what is felt for the living being. The cogitative sense relates what the living being knows to its situation and interests, providing a certain subjective perspective to what is felt. Through this sense, sensitive knowledge connects with desire and behavior. Memory also has a subjective component, since it retains sensations and evaluations with the connotation of having being lived. Through memory, a sensation becomes an experience, and it thus provides the association with the past that is the basis for any experience of one's own subjectivity. It is through memory and the cogitative sense that subjectivity appears in sensibility.

The cogitative sense generates sensory expectations, but these belong to us and are not features of the external world, while memory gives us a perspective of the past, that which has happened to us. But while all of these capacities are exercised through the body, it is not clear how the body can appear to them in a unitary way. These two internal senses are oriented primarily towards behavior, not the living being's contemplation of itself. Expectations and experiences are linked to the body in their incidence and their realization, but are not primary and explicitly about the body per se. For this reason, Aristotelian psychology, particularly its theory of internal sensitivity, does not seem to offer grounds for a theory of the body in the phenomenological sense.

However, if we accept the Aristotelian theory of knowledge, sensitivity at all levels is knowledge in the strict sense, but it is not yet intellectual knowledge. The intellect – the *nous* – is distinguished because it grasps reality as it is. Only the *nous* allows knowledge of the truth, thereby permitting the development of science and, with it, true knowledge.

To perceive movement as such it is necessary that there be a point of view above or outside movement and this is what we call presence. Like Aristotle, Leonardo

[15] The notion of the cogitative sense does not appear as such in Aristotle, but is common in the later Aristotelian tradition.

Polo, who saw his work as an exposition and continuation of Aristotle's philosophy, does not see the present as a dimension of time that can be distinguished from future and past independently of subjectivity. For Polo, presence is a human intellectual operation, and the present is its intentional object, the point of reference that allows for the temporal articulation of sensitivity, past experience, and memory – in Aristotelian terms – as well as expectations, i.e. sensory valuations that the cogitative sense carries out. Only in the present, from a position removed from movement, but about which movement revolves, can movement be understood and measured.

Presence is the first level of human intellectual activity, and "making present" is the most basic feature of thinking. Only when we introduce presence can we have knowledge in terms of temporality. Only from a point of view external to movement can one understand movement as the real distinction between "before" and "after." Without this activity, which grasps things without motion and is foreign to physical movement, we would live in movement (*vita in motu*), but our existence would not be properly defined by temporality, the consciousness of time.

7.8 Presence, Subject, Self

On the basis of previous discussion, what is the relationship between that which is present and the self? One of the problems of modern philosophy lies in its careless identification of presence and subjectivity. The subject, we can say, is often conceived as "attending" in the present to the passage of time, and absolute subjectivity – if one accepts such a thing from an idealistic perspective – is seen as the capture of the entire unfolding of reality in the present. When the mind-body problem arises, this is usually the way that self-consciousness and subjectivity are conceived.

However, it turns out to be impossible to combine a presence – or a supposed self-presence – with a body, whether we conceive it as a mere *res extensa*, or affirm its dynamic and elusive character.

In my opinion, the root of this difficulty is the failure to realize that the present cannot be dealt with in the same way as other objects. It is introduced by an activity, which is an essential feature of our intellectual condition: the activity of thinking. This activity of the living being is precisely the reality of presence. Presence is the act of presenting, and in so doing it separates and considers formal content from a totally akinetic perspective. This activity makes possible the consideration of movement as a process or *kínesis*, and makes it possible to compare it in an objective way with other movements, namely by measuring it.

Leonardo Polo has also indicated that presence does not exhaust human intellectual knowledge. The very possibility of knowing presence as an "operation"[16] (*praxis*) shows that knowledge goes beyond it. In fact, the human self manifests

[16] The word *enérgeia* which I have translated in this text as "activity" is usually translated into Latin as *actus*.

itself as having more strength in its practical action than in its grasping of theoretical knowledge (Murillo 2009). In this case, though, the self cannot be confined to the present. Action is always open to the future, which, incidentally, is much more than the "after." In the realm of action it is evident that the self is not something given once and for all, but also has a character that involves demands and aims – projects. I believe that this is the key to understanding the importance Aristotle attaches to happiness (*eudaimonia*) as the horizon of authentically human action, which is a consequence of the orientation of the self towards the future.

Specifically, intellectual activity, which enables us to live the present, is sui generis compared to other vital activities, as Aristotle showed. Among other things, as Thomas Aquinas emphasized, thinking differs from sensory activities: although thinking needs the body to be exercised in order to take the senses as its object, the faculty that exercises it cannot be attributed properly to the activity of any organ (Aquinas 1961, II, 66).

However, this distinction between sensory activities, which are corporeal, and intellectual ones, which are not corporeal in the same way, does not imply a separation between two things or types of substances. The Aristotelian conception holds that life exists in and through movement (*vita in motu*). The living being is not a system or a process, but a coordination of activities, some of which are physically realized as movements and can be scientifically described as processes.[17] This fact makes feasible a naturalistic approach to the study of life, as is the case with the consideration of organisms as systems and of vital activities as processes. However, considered in themselves, these movements intrinsically differ from others in that they are possessions of an end, and can only be understood as "vital" to the extent that one understands that they are aimed primarily at the continuation and improvement of the life of the being and not at producing a result external to themselves. Vital activities are the expression and condition of possibility of the living being, whose life is also an activity.

What is the relationship of presence to the body? We have already seen that for the sensitive faculty, the body, as represented by the sensory organs, remains hidden. The subjective experience of my corporeality is often linked to a particular kind of sensation, like pleasure or pain. These afford qualitative information, which is susceptible to varying degrees of intensity. Of the two, pain seems to be more linked to the awareness that we have a body. Tactile pleasure can be located, but it seems to involve a certain emancipation from the limits of the body. Pain, however, is often the most unmistakable way by which we perceive that we have a body, and distinguishes it from "what is not us." Although I might use a stick as an extension of my body, the point at which pain begins serves as a criterion to distinguish between the stick and the body. The consciousness of corporeality, then, is closely united to awareness of boundaries. While pleasure is, as Aristotle put it, unimpeded natural activity (Aristotle 2002, VII, 1153 a 13–16), it seems instead to transport us beyond itself. Perhaps for

[17]This fact legitimates the consideration of living beings as systems and their activities as processes, provided that they are not ontologically characterized by these concepts.

this reason, action, which also transports us outside of the body, is pleasant in itself, although it may incidentally turn out to be tiring or painful.

However, while we can have an experience of the body, and although the body is required for thinking and introducing presence, the absence of the body while thinking appears at this level as an enigma: the correspondence between body and thinking cannot be discovered by thinking, although this correspondence is undeniable (Polo 2003, p. 280). Polo indicates that the correspondence of thinking with one's own body is one of the meanings of facticity. One's own body is the un-thought fact, which corresponds to the indiscernibility of thinking and not thinking (2003, pp. 279–80).

In this understanding, the body does not appear as something ready-made, with which the self or thought must be combined. Its vital meaning is not that of an extended substance that the self can control, but rather that it is the basis for life and the vital activities that impel and configure it. In the human being's case, amongst these vital activities, thinking introduces the present, and with it the problem of the factuality of the body and that of the body's own identity over time.

I return, then, to our initial question: am I my body? If we understand the body as something already given, as a mere product, we are in fact providing a description of my corpse – what remains of me after my death, i.e. that which is no longer me – rather than of me. But, if we do not separate the body from movement and the activities through which it exists and is alive, this identification is to some extent possible. However, it is not enough to say that the self is the body, because the activities by which I perceive myself as a self are not, strictly speaking, corporeal. Neither is it entirely correct to say that I have a body. The body is instead the condition of possibility and the vehicle of possession: I have because I have a body, that is, because I possess by way of activities exercised through the body.

Thus, the consideration of the body not as a thing but as a condition and result of certain movements and activities sheds a different light on the problem of dualism. The body can no longer be regarded as something irrelevant to the activities of the living being. This allows us to include other activities as dimensions of certain living beings, such as thinking, acting, etc., which share the character of not being a "thing" with the other activities of the living being. However, without these activities the body observable from the outside would not be what it is: the external manifestation of a living being.

References

Aquinas, Th. 1888–1889. *Pars Prima Summae Theologiae*. Opera omnia, IV–V. Rome: Ex Typographia Polyglotta S. C. de Propaganda Fide.

Aquinas, Th. 1961. In *Summa Contra Gentiles, Books 2–3*, ed. P. Marc, C. Pera, and P. Caramello. Turin/Rome: Marietti.

Aristotle. 1924. *Metaphysics*. W.D. Ross, 2 vols. Oxford: Clarendon Press. Reprinted 1953 with corrections.

Aristotle. 1983. *Physics. Books III and IV*, trans. with notes E. Hussey. New York: Oxford University Press.

Aristotle. 2002. *Nicomachean ethics*. Trans. (with historical introduction) Ch. Rowe; philosophical introduction and commentary S Broadie. New York: Oxford University Press.

Behnke, E. 2004. Edmund Husserl's contribution to phenomenology of the body in *Ideas II*. In *Phenomenology. critical concepts in philosophy. Volume II: Phenomenology: Themes and issues*, ed. D. Moran and L.E. Embree. New York: Routledge.

Carman, T. 1999. The body in Husserl and Merleau-Ponty. *Philosophical Topics* 27(2): 205–226.

Craig, A.D. 2003. Interoception: The sense of the physiological condition of the body. *Current Opinion in Neurobiology* 13(4): 500–505.

Fuchs, Th. 2000. *Leib, Raum, Person. Entwurf einer phänomenologischen Anthropologie*. Stuttgart: Klett-Cotta.

Fuchs, Th. 2013. *Das Gehirn – ein Beziehungsorgan: eine phenomenologische und ökologische Konzeption*, 4th edn. Stuttgart: Kohlhammer.

Hegel, G.W.F. 1992. *Gesammelte Werke. 20, Enzyklopädie der philosophischen Wissenschaften im Grundrisse, (1830)*. Hamburg: Felix Meiner.

Heidegger, M. 1996. *Being and Time*. Trans. J. Stambaugh. Albany: State University of New York Press.

Husserl, E. 1984. *The Crisis of European Sciences and Transcendental Phenomenology: An Introduction to Phenomenological Philosophy*, trans. D. Carr. Evanston: Northwestern University Press

Husserl, E. 1989. *Ideas pertaining to a pure phenomenology and to a phenomenological philosophy. Second Book. Studies in the phenomenology of constitution.* Trans. R. Rojcewicz and A. Schuwer. The Hague: Martinus Nijhoff.

Jonas, H. 1992. Evolution und Freiheit. In *Philosophische Untersuchungen und metaphysische Vermutungen*. Frankfurt: Insel Verlag.

Jonas, H. 2001. *The phenomenon of life: Toward a philosophical biology*. Evanston: Northwestern University Press.

Kant, I. 1998. *Critique of pure reason*. Trans. A. Wood and P. Guyer. New York: Cambridge University Press.

Leder, D. 1990. *The absent body*. Chicago: University of Chicago Press.

Merleau-Ponty, M. 1964. The philosopher and his shadow. In *Signs*. Trans. R. McCleary. Evanston: Northwestern University Press.

Merleau-Ponty, M. 2010. *Oeuvres*. Paris: Gallimard.

Metzinger, Th. 2003. *Being no one. The self-model theory of subjectivity*. Cambridge, MA: MIT Press.

Murillo, J.I. 1999. ¿Son realmente autónomas las ciencias? In *Fe y Razón. I Simposio Internacional Fe cristiana y cultura contemporánea*, ed. J. Aranguren, J. Borobia, and M. Lluch. Pamplona: Eunsa.

Murillo, J.I. 2008. Health as a norm and principle of intelligibility. In *Natural law: Historical, systematic and juridical approaches*, ed. A.N. García, M. Silar, and J.M. Torralba. Newcastle: Cambridge Scholar Publishing.

Murillo, J.I. 2009. ¿Comprender la libertad? Entre la biología y la metafísica. *Anuario Filosófico*, XLII/2: 391–418.

Murillo, J.I. 2011. Apariencia y realidad del yo: una aproximación crítica a la propuesta de Thomas Metzinger. In *Asalto a lo mental. Neurociencias, consciencia y libertad*, ed. F. Rodríguez Valls, C. Diosdado, and J. Arana. Madrid: Biblioteca Nueva.

Plessner, H. 1981. *Gesammelte Schriften. IV: Die Stufen des Organischen und der Mensch*. Frankfurt: Suhrkamp.

Polo, L. 2003. *Antropología trascendental. Tomo II: La esencia de la persona humana*. Pamplona: Eunsa.

Polo, L. 2006. *Curso de teoría del conocimiento. Tomo I*. Pamplona: Eunsa.

Saborido, Ch, M. Mossio, and A. Moreno. 2011. Biological organization and cross-generation functions. *British Journal for the Philosophy of Science* 62: 583–606.

Scheler, M. 1973. *Formalism in ethics and non-formal ethics of values. A new attempt toward the Foundation of an Ethical Personalism.* Trans. M.S. Frings and R.L. Funk. Evanston: Northwestern University Press.

von Schelling, F.W.J. 2000. *System des Transzendentalen Idealismus*. Hamburg: Meiner.

Schmitz, H., R.O. Müllan, and J. Slaby. 2011. Emotions outside the box—The new phenomenology of feeling and corporeality. *Phenomenology and the Cognitive Sciences* 10(2): 241–259.

Sider, T. 2001. *Four-dimensionalism*. Oxford: Clarendon.

Snell, B. 2000. *Die Entdeckung des Geistes: Studien zur Entstehung des europäischen Denkens bei den Griechen*. Göttingen: Vadenhoeck und Ruprecht.

Sokolowski, R. 2008. *Phenomenology of the human person*. New York: Cambridge University Press.

Spaemann, R., and R. Löw. 2005. *Natürliche Ziele: Geschichte und Wiederentdeckung des teologischen Denkens*. Stuttgart: Klett-Cotta.

von Bertalanffy, L. 1969. *General system theory: Foundations, development, applications*. New York: George Braziller.

Welton, D. 1999. *The body: Classic and contemporary readings*. Malden: Blackwell Publishers.

Chapter 8
The Enactive Philosophy of Embodiment: From Biological Foundations of Agency to the Phenomenology of Subjectivity

Mog Stapleton and Tom Froese

8.1 Introduction

Following on from the philosophy of embodiment by Merleau-Ponty, Jonas and others, enactivism is a pivot point from which various areas of science can be brought into a fruitful dialogue about the nature of subjectivity. In this chapter we present the enactive conception of agency, which, in contrast to current mainstream theories of agency, is deeply and strongly embodied. In line with this thinking we argue that anything that ought to be considered a genuine agent is a biologically embodied (even if distributed) agent, and that this embodiment must be affectively lived. However, we also consider that such an affective agent is not necessarily also an agent imbued with an explicit sense of subjectivity. To support this contention we outline the interoceptive foundation of basic agency and argue that there is a qualitative difference in the phenomenology of agency when it is instantiated in organisms which, due to their complexity and size, require a nervous system to underpin their physiological and sensorimotor processes. We argue that this interoceptively grounded agency not only entails affectivity but also forms the necessary basis for subjectivity.

M. Stapleton (✉)
Institut für Philosophie, Universität Stuttgart, Stuttgart, Germany

Philosophy of Neuroscience (PONS) Group, Centre for Integrative Neuroscience, University of Tuebingen, Tuebingen, Germany
e-mail: mog.stapleton.philosophy@gmail.com

T. Froese
Instituto de Investigaciones en Matemáticas Aplicadas y en Sistemas, Universidad Nacional Autónoma de México, Mexico City, Mexico

Centro de Ciencias de la Complejidad, Universidad Nacional Autónoma de México, Mexico City, Mexico
e-mail: t.froese@gmail.com

M. García-Valdecasas et al. (eds.), *Biology and Subjectivity*, Historical-Analytical Studies on Nature, Mind and Action 2,
DOI 10.1007/978-3-319-30502-8_8

To begin with, we introduce an emerging movement in cognitive science and related fields, known as enactivism. The enactive approach to cognitive science brings together several fields of study into one coherent research program of life, mind and sociality (Thompson 2007; Di Paolo and Thompson 2014). It thereby inherits the interdisciplinary perspective that is characteristic of the cognitive sciences, but puts special emphasis on a number of additional fields that have been neglected by the mainstream. In particular, the enactive approach stands out by bringing together two venerable traditions of continental philosophy, Husserlian phenomenology and the philosophy of the organism, with cutting edge research in the sciences of complexity, as formalized by dynamical systems theory (Weber and Varela 2002). Perhaps surprisingly, the insights gained by phenomenology lend themselves to being described as structures in a temporal flow using dynamical systems theory (Varela 1999), and Husserl's method of using imaginative variation to reveal a phenomena's essential characteristics is not far removed from computer-aided systems modeling of minimal cognition (Froese and Gallagher 2010). Similarly, key philosophical claims about the self-organizing and self-producing nature of organisms, going back at least to Kant, are starting to find expression in the fields of AI, systems biology and artificial life modeling (Froese and Ziemke 2009; Di Paolo 2010).

This confluence of approaches puts enactivism in a privileged position to investigate the relationship between biological embodiment and phenomenological subjectivity (e.g. Desmidt et al. 2014). It does so by conceptualizing the objectively living body (*Körper*) and the subjectively lived body (*Leib*), following Husserl's ([1952] 1989)[1] terminology, as two sides of the same coin. The traditional mind-body problem is therefore converted into the more tractable "mind-body-body problem" (Hanna and Thompson 2003). And we are given an additional mediating concept that has been almost completely neglected in analytical philosophy of mind, but which has taken center stage in enactive theory, namely life itself:

> The Mind-Body-Body Problem […] is how to understand the relation between (i) one's subjective consciousness, (ii) one's living and lived body (*Leib*), that is, one's animate body with its "inner life" and "point of view;" and (iii) one's body (*Körper*) considered as an objective thing of nature, something investigated from the theoretical and experimental perspective of natural science (physics, chemistry, and biology). (Hanna and Thompson 2003)

This constellation of phenomena has long fascinated the phenomenological tradition of philosophy. For example, Scheler ([1928] 2008), Plessner ([1928] 1975), Merleau-Ponty ([1942] 1983), Sartre ([1960] 2004) and Jonas ([1966] 2001) were all in their own ways interested in grounding the origins of human subjectivity in the most basic principles of organic being and behavior. To be sure, accepting that the human condition is partially constituted by forms of organic and animal life does not entail that these forms explain all there is to being human. For example, Heidegger ([1929] 1995) was inspired by the biologist von Uexküll (1909, [1934]

[1] Stylistic note: we cite the most recent English edition of our sources whenever this is possible for ease of reference. However, in order to avoid giving a false impression of the original date of publication, we also always provide the year of publication of the first edition in square brackets.

1957), who argued for the existence of an organism's own point of view (i.e. its *Umwelt*), to contrast the restricted point of view of the animal from the conceptual world of the human. Nevertheless, Heidegger recognized that our being-in-the-world (*Dasein*) depends both on *Dasein*'s understanding of being as well as on *Dasein*'s living embodiment (Kessel 2011).

This tradition of relating subjectivity with biological embodiment is not a mere historical relic; it is continued by modern phenomenological philosophers such as Barbaras (2005), Gallagher (2005) and Zahavi (1999), who also engage with the ongoing development of enactive theory (e.g., Barbaras 2002, 2010; Zahavi 2011; Gallagher 2012). In the following we complement these efforts by sketching an enactive approach to the question of how a physical living body (*Körper*) can be an affectively lived body (*Leib*) and also a reflectively lived consciousness. Specifically, first we will describe what kind of embodiment is an essential prerequisite for affective agency and then we will consider some additional biological constraints that are imposed by phenomenological subjectivity. Here we are not concerned with giving the biological or the phenomenological domain metaphysical priority but, following Varela's (1996) working hypothesis of neurophenomenology, with putting insights from both domains into a relationship of mutual constraints to further our understanding of the phenomenon of life as a whole.

8.2 Biological Foundations of Agency

What is it about biological cognitive systems that makes us want to talk in terms of them being agents? It cannot merely be that they move around in and make changes to the environment – after all a robotic vacuum cleaner achieves this very effectively. If we admit such a reactive robot into the class of agents, then we have to include the humble thermostat as well, and the concept of agency eventually ends up being so inclusive so as to be theoretically useless (Froese 2014). Nor can it be that a system must have a thematised *feeling* of agency, explicitly experiencing themselves as the author of their actions to the point of reflective self-awareness. Such a strong stipulation would plausibly rule out most non-human animals and quite possibly even infants: biological systems which – even though they may not be aware of it – intuitively instantiate some sort of basic subjecthood beyond its mere attribution by others.[2] There is something deep in the notion of agency, such that we know that merely attributing agency to a system, from our perspective as external observers, is not sufficient for it to genuinely be an agent (Rohde and Stewart 2008). Agency is intrinsic to the system itself, but in virtue of what?

The most obvious feature of systems which one might consider as uncontroversially agentive is that they have wants and needs. The kinds of robots which are

[2]The attribution of subjecthood by others may in fact be an essential element in infants' development of an explicit awareness of their own subjectivity (Reddy 2003), but on the view we are promoting here they were already agents on their own terms even before they were born. This is what allows much of the body schema to develop *in utero* rather than post-partum (Lymer 2011).

commonly termed 'agents' in informatics seem – at least on the surface – to have these, typically because they have been programmed, designed or evolved so as to move as if they are satisfying needs (e.g. Parisi 2004). Biological systems, of course, do not need to act *as if* they have them. As Weber and Varela (2002), following Jonas ([1966] 2001), argue, their needs are fundamental to their continued existence: a living system that no longer has any needs to satisfy is in fact no longer a living system. As an aside, we note that the phenomenon of an agent's wants is a bit more difficult to address, because fulfilling a desire goes beyond the mere satisfaction of an existing, well-defined lack to an open-ended quest for something that is not yet present (Barbaras 2002, 2010). As such, desires are probably dependent on a more explicit awareness of one's own subjectivity.

The enactive theory of agency has its roots in these biological foundations of cognitive systems. A precursor to its theoretical framework started taking shape in the cybernetics movement (Froese 2010), for example in attempts to explain organisms' purposeful behavior in terms of feedback dynamics. Another important milestone was Maturana and Varela's (1987) work on the self-organising and self-producing properties of minimal living systems such as the cell. They argued that the cell is the minimal living system because it forms itself as an identity – that is to say it forms itself as a system distinct from its environment. Its particular organization allows it to be self-producing – the processes that go on inside the cell produce the boundary of the cell which distinguishes it as an identity, and this boundary allows these processes within the cell to keep going and producing it – a circularly causal process (Varela et al. 1974; Varela 1979). This form of organization of the living system was termed *autopoiesis*, a concept that is closely related to a number of other technical concepts.

While autopoietic systems are material systems, what is key to the formation of their own systemic identity is the specific organizational nature of the metabolic processes rather than the particular material processes with which that organization is realized (for an introductory overview, see Di Paolo and Thompson 2014). The autopoietic organization is defined as autonomous because it has the property of operational closure, which means that the organization subserves a dynamic process of self-generation under far-from-equilibrium conditions. In other words, an autonomous system is organized as a network of processes that mutually depend on each other, and on the organization of the whole network, for their continued existence. Although there are similarities between these ideas and the cybernetics of feedback systems, the crucial difference is that autonomy ensures that the system is genuinely self-determining from the bottom up and not just self-maintaining a set of externally defined conditions (Froese and Stewart 2010).

These requirements are not trivial, but nevertheless an autopoietic system could spontaneously emerge and continue to exist and self-generate without having to be a sensorimotor system – providing that it exists in optimal conditions that provide everything it needs for its continued self-production. In addition, there are some interesting examples of self-organizing material systems that may or may not be autopoietic systems, such as a tornado or dust devil. Here we need to be careful to distinguish self-maintenance from self-production. Autopoiesis refers to a network of processes of *production* of new components, whereas a tornado and a dust devil presumably are only *rearranging* pre-existing components. Things are not quite as simple as this (see, e.g., the critical discussion by McGregor and Virgo 2011), but the difference between synthesis of new components compared to mere re-organization of existing components provides a useful heuristic.

Nevertheless, a spontaneously emerging autopoietic system – even if it were autonomous, would not be *agentive* – its movements and indeed entire existence are at the mercy of external factors and its survival as a system is just a matter of luck that the right conditions for its continued existence happen to occur. In actual fact, even the simplest autopoietic systems, such as reaction-diffusion systems, are not *entirely* passive; they are capable of self-movement and a limited range of interactions with their environment (Froese et al. 2014). But the important point here is that these basic autopoietic systems are not capable of actively regulating their behavior in relation to their needs. Although in the early formulations of the enactive approach an equivalence between autopoietic, living, and cognitive systems was assumed (e.g., Stewart 1992), this has started to be questioned. Some enactivists currently argue that autopoiesis is necessary but not sufficient for life and life necessary but not sufficient for cognition (Froese and Di Paolo 2011) while others argue that it is *autonomy* rather than autopoiesis that is necessary for cognition (Thompson 2007; Thompson and Stapleton 2009). The precise relation between autopoiesis, autonomy, and life remains an open question for future research (see in particular the discussions in Thompson 2011; Wheeler 2011).

Di Paolo (2005) enhanced the concept of autopoiesis with that of adaptivity in order to yield an organizational structure which subserves the kind of systems that we consider to be agents. An autopoietic system, however it is instantiated, is going to have limits to what kind of changes can happen in the environment and within its own systems that still allow the system to continue. At some point, however robust the system is, if the changes from the organization are too great it is not going to be able to self-produce and self-maintain. The set of changes that can happen within these limits are its viability set. According to Di Paolo (2005), in order for an autopoietic system to be able to continue its existence under changes in the environment, rather than just cease to exist as a system, it needs to "(i) be capable of determining how the ongoing structural changes are shaping its trajectory within the viability set, and (ii) have the capacity to regulate the conditions of this trajectory appropriately" (Froese and Di Paolo 2011, p. 8). This is the property of *adaptivity* (Di Paolo 2005; see also Barandiaran and Moreno 2008). There is a growing consensus in enactivism that autopoiesis and adaptivity are necessary and sufficient for life, and that therefore living is sense making because the underlying adaptive processes are normative (Thompson 2011).

The processes necessary for adaptivity can occur within the system, i.e. by means of modification of the internal milieu, but only to a certain extent. In order to be more adaptive, a system must be able to adjust its relationship with its environment, such as by moving its position to a more favorable location, in order to change the effect this environment has on its viability. And because these changes in its environment caused by its moving are related to what the system needs in order to continue its existence, this movement is not a mere passive motion but is realized according to some goal or norm (survival), and is thus defined as an action. This gives the system intrinsic agency.

Developing this idea, Barandiaran et al. (2009) propose an operational definition of agency. After critically reviewing existing definitions of agency, they argue that "we can generalize that agency involves, at least, *a system doing something by itself according to certain goals or norms within a specific environment*" (p. 369), and flesh this out with the necessary and sufficient conditions/aspects of agency, namely individuality, interactional asymmetry, and normativity. More specifically,

> (i) there is a system as a *distinguishable entity* that is different from its environment, (ii) this system is *doing* something by itself in that environment and (iii) it does so according to a certain goal or *norm*. (Barandiaran et al. 2009, p. 369).

Let us consider these three requirements of agency in more detail.

(1) **Individuality:** Individuality needs to arise from the structure of the system itself rather than be attributed from an outside observer. Thus an artificial system could conform to this requirement, but only if the processes which constitute it as a system bind it together in a coherent way and by doing so distinguish it from the environment with which it interacts (Froese and Ziemke 2009). It does not suffice to just distinguish a part of the world and view it as a system for our explanatory purposes. It may be useful to think of some non-self-individuated parts of the world as active systems, but to do so is to think of them from *our perspective* and not as a result of any particular dynamics of the system in question *as an individual*. To think in this way therefore does not imply that such systems have an identity of their own (Froese 2014). A mobile robotic 'agent' of the kind standardly found in robotics labs therefore does not fulfill this criterion – it is we who demarcate the mechanisms of the robot as the relevant system for our explanatory project. We could just as well demarcate the level of its interactions with other robots as the level of the system – as we do in swarm robotics. Or even 'agent + environment' as the system (Beer 2000). It all comes down to what our explanatory project is. There does not seem to be a basis for a more or less 'correct' attribution of the system's boundaries. Compare this to a living cell. In the case of the cell the tight interactions between its components feed back to each other and enable each other to exist and to continue as such, i.e. the system consists of a network of operationally closed processes. It is – ontologically – an individual because when we, in our role as external observers, distinguish it as a system appropriately, then the system will reveal itself to us as actually autonomously distinguishing *itself*, that is, as existing independently of our epistemological choices and distinctions.

(2) **Interactional asymmetry:** for agency it is not sufficient for an individual system to just be a moving system, nor to merely be in interaction with the environment or other systems. Nor is it sufficient for it to rely on a subsystem (which is not relevantly interconnected with the rest of the system) that drives its movement (e.g., the idea of metabolism-independent chemotaxis, see Egbert et al. 2012). The movement must arise at the agent level as a whole so that the agent

uses the movement to modulate its coupling with the environment. The key point here is that agency requires that the adaptive regulation of agent-environment interaction is realized by the agent rather than resulting only from contributions of the environment. This distinction is important because some types of behavior can be adaptive at the level of the population, i.e. they lead to enhanced reproductive success, while still remaining reactive at the level of the individual, i.e. they are driven mostly by the environment (Froese, et al. 2014). Active regulation of interaction does not need to be happening all of the time in order for us to attribute agenthood but it must be a relevant aspect of the system. It is this aspect that underpins the system's greater adaptability as its regulated movement allows it to find an environment that better suits its viability set.

(3) **Normativity:** The concept of normativity should not be misunderstood as applying only in relation to human cultural conventions that guide action, although in that context it certainly finds a particularly elaborated expression (Torrance and Froese 2011). What is of interest here is something more fundamental, namely the biological norms that guide adaptive behavior. For a movement to be an action of an agent it must be a movement of an individual, have interactional asymmetry (arise from the agent modulating its coupling) *and* be relevant to some goal – a goal which it either achieves or fails to achieve. This goal should not be externally given and the system arbitrarily directed toward its realization, such as when designing an artifact to behave in certain ways. Rather the goal should arise from and be relevant to the system's self-producing and self-maintaining activity. In a manner of speaking the system must aim at a goal in order for its movement to be an action, and it must be possible to fail at achieving this goal. However, we should beware of letting our manner of speaking mislead us into reifying the basis of this biological normativity into hypothetical entities of some sort, such as explicit representations of goals and norms. These are clearly operative in the cultural domain, for example in the legal system (Gallagher 2013), but they get in the way of operational explanations aimed at the subpersonal level (Di Paolo et al. 2010). Such goals or norms emerge within the living system as a result of the autonomous or adaptive dynamics (metabolic or otherwise) attempting to keep the system within its boundaries of viability (Barandiaran and Egbert 2014).

An account of agency that requires individuality, interactional asymmetry, and normativity gets us some of the aspects which are key to our intuitive notion of agency, in particular its being the source of action and intrinsic intentionality. While these requirements may not be sufficient for realizing agency in all its forms, they seem to at least be instantiated in systems to which we intuitively do attribute agency. And, of particular import, such an account is operationalisable and thus, even though it is grounded in biology, it is not biologically chauvinist. To say that an account of agency is 'not biologically chauvinist' is to say that it does not rule out the possibility of a non-biological instantiation of agency (see also the discussion by Thompson 2011). The enactive conception of agency – grounded in the systemic principles of autonomy and adaptivity – is nevertheless embodied in a non-trivial way, because it is not only that the body instantiates properties that make it an agentive system but that these functions are by their very nature grounded in biology: the biology of value. What this means is that even if it is operationalisable and realizable in an artificial system, this artificial system will be, in a sense, an artificial biological system because the values that arise out of the system and hence its normativity are grounded in its autonomous (in this case *artificial-autopoietic*) self-production and adaptive self-maintenance, hence it would match our criteria of a living system – even if it is not composed of typical organic material.

Agency is one of the ways in which enactivism is in tension with orthodox (functionalist) philosophy of mind and cognition. If the agential system were a purely functionalist system then it ought to be realizable in any kind of different "hardware". Yet if values arise from the self-creation and self-maintenance of a particular system, then if you abstract that system from the physical basis that it creates and with which it maintains itself, then those values cease to exist for that system (Di Paolo 2009). This might seem at odds with the fact that our definition of agency is based on organizational criteria and therefore ought to be able to be instantiated in a variety of different systems – not only living ones (for a more detailed discussion of related concerns, see Wheeler 2011 and Thompson 2011). However, while the organization of such an agent is indeed relatively independent with respect to its particular physical realization, this does not mean that anything goes, as would be the case for functionalism. For example, the essential role of mortality in meaning-generation entails that agency cannot be completely divorced from a precarious existence in some material substrate (Di Paolo 2009; Froese in preparation), and this necessity of far-from-equilibrium self-production and self-maintenance imposes even more specific material constraints. For example, the material of the components of the system cannot be inert, but robots are typically built from such material (Moreno and Etxeberria 2005). And it is the particular material instantiation we suggest – what Stapleton (2012) dubs a system's "particular embodiment" – that gives rise to the particular values and norms of that system.

Advocating that it is a system's particular embodiment that matters for agency, it should be noted, is not *in principle* at odds with the functionalist approach to embodiment. For example in Clark's (1989) work on "microfunctionalism" he argues that functionalism need not be identified with formal descriptions pitched at a gross level, but that what is essential to functionalism is merely that the "structure, not the stuff, counts" (Clark 1989, p. 31). In a similar vein Wheeler, drawing on the University of Sussex evolutionary hardware and robotics paradigm of A. Thompson (e.g. Thompson 1995, develops Clark's (1997) notion of "continuous reciprocal causation" arguing that in evolved systems (both biological and artificial) the "low-level" properties of the hardware are relevant to adaptive success (Wheeler 2005, pp. 267–68)). For these reasons at least one of us (Stapleton) is inclined towards what she calls a "nanofunctionalist" paradigm (see Stapleton 2012, Chapter 5 and Conclusion). This is the position that the relevant level for understanding cognition in natural cognitive systems is very close to (and in some cases entwined with) the implementational level. Nevertheless, what is important about this implementation is the (nano-)functional role it plays for the system. Such a position however runs the risk of irritating both traditional (functionalist) embodiment theorists who have explicitly rejected radical embodiment (see for example Clark 1999; Wheeler 2010) for not abstracting from implementation enough, and enactivists who reject the functionalist tradition for abstracting from implementation too much.

It is however indeed the case for enactivists that what we intuitively consider as cognition involves a certain amount of autonomy with regard to its bodily basis, for example the relative operational autonomy of the nervous system (Barandiaran and Moreno 2006). Another mark of the cognitive may be that it serves to decouple an agent from its environment by mediating its sensorimotor interactions (Fuchs 2011). This decoupling can be achieved in a variety of manners, and enactivism is currently exploring how such mediation can help to bridge the cognitive gap from basic agency to more full-blown forms of human action (see review by Froese 2012).

8.3 From Affective Agency to Subjective Self

Given that the enactive conception of agency already involves an appeal to the normativity of adaptive behavior, it also provides a useful foundation for grounding the affective dimension of animal existence. Colombetti (2010), following Weber and Varela (2002), argues that the processes which subserve the biological (autopoietic) organization of living systems not only establish a "point of view" and locus of agency but also the enaction of meaning. The idea is that those parts of the world that are relevant to the self-production and self-maintenance of a system have meaning for the system. Here we are not concerned with linguistic semantics. Meaning is another aspect of the values and norms discussed earlier: all are generated from within the system as a result of its relation to those parts of the world it interacts with (i.e. its "Umwelt", according to von Uexküll's terminology). For something to be meaningful to the system is for it to have value for it, and thus for it to have a normative character. Colombetti notes that Di Paolo's (2005) addition of adaptivity to the previously binary (alive or dead) notion of autopoiesis allows us to account for the grades of meaning (or "degrees of value") offered by a system's environment, and thereby "makes room for a notion of organismic preferences" (2010, p. 149).

When we understand cognition in these terms, the self-regulatory metabolic (homeostatic) functions from which the values and meaning of a particular living system emerge are also those that ground emotion for neurobiological theorists such as Damasio (1999, 2010) and Panksepp (1998). Colombetti therefore concludes that "[o]n this view of emotion, the account of natural purposes developed by Weber and Varela (2002) and Di Paolo (2005) as a theory of bodily sense-making is as much a theory of emotion as it is a theory of cognition" (2010, p. 150). In general, enactivism provides a fitting framework for the tight integration of emotion, cognition and perception (e.g. Colombetti 2007, 2014; Colombetti and Thompson 2008; Thompson and Stapleton 2009; Varela and Depraz 2005; Ward and Stapleton 2012; Bower and Gallagher 2013).

This integration of affect and agency can be seen clearly when we consider how value and action are integrated in organismic systems. In a very simple system such as a single cell the internal metabolic dynamics and those underpinning the cell's sensorimotor functions are not segregated very well – although precisely how independent they are from each other is debatable (Egbert et al. 2012). Nevertheless the internal value – from the physiological condition of the cell – is entwined with the action of the cell to maintain itself within its viability set. So even though it may not be a consciously feeling system, it is nevertheless right to think of it as an affective system: the world for the cell is shaped by this affect – it is what imbues the world with value for it, and thus what imbues the cell with normativity.

Once the agentive system is constituted by multiple cells, two issues arise: the complexity of the modulations required, which depend on a larger ensemble of physiological states, and the time it takes for processes in one part of the system to affect more distal processes. These increases in internal spatiotemporal distances and their bridging in terms of increased bodily sensing and regulation go hand in hand with an increase in decoupling from the environment, or at least an expansion of the role of self-mediation in action and feeling. We are therefore approaching the transition from mere affective agency to full-blown subjectivity. In the following we consider essential aspects of the internal organization of this more specific form of embodiment in some detail.

What is needed for multicellular systems then is some sort of homeostatic mechanism, which is both sensitive to the animal's internal milieu and able to regulate it effectively, while at the same time providing the basis for adaptive interaction with the world. In animals these functions are provided by the nervous system. While typically the term 'interoception' is used to refer to the sensitivity or awareness of internal, visceral changes as mediated by the autonomic part of the peripheral nervous system, we suggest that this is a contingent fact based on the predominance of research into humans. Interoception as the sense of the internal body after all, may be mediated through molecular communication networks in, for example, the endocrine and immune systems (Cameron 2001). That complex multicellular creatures like us require a functioning interoceptive nervous system as well does not detract from the basic structure of interoception being essentially a sensitivity to internal modulation which is needed in order to effect internal and external changes for the purposes of maintaining or increasing the adaptivity of the system. While this may just be another way of specifying the internal dynamics of autonomous, adaptive systems, we believe that understanding the mechanisms by which these organisational criteria are brought about in biological systems will yield an increased understanding of the requirements for more complex forms of agentive systems in general.

Two questions in particular seem to fall out of this. Firstly, how is interoception realized in humans and can this specific mechanism be operationalized beyond our particular embodiment? And secondly, how does the instantiation of an interoceptive "nervous" system in an agentive system qualitatively alter the system in terms of its agency? Finding answers to these questions is crucial for developing an enactive approach to agency that goes beyond an account of organismic agency in general, toward an account of animality more specifically and, most importantly, of human subjectivity. In the following we offer some considerations about how to begin answering this open challenge.

The term 'interoception' was introduced by Sherrington in the early twentieth century and used by him to distinguish the sense of the visceral body from the sense of limb position (proprioception), the senses of touch, pain, and temperature (exteroception), sight and hearing (telereception) and taste and smell (chemoreception) (Sherrington 1948, cited in Craig 2002). Nowadays vision, hearing, smell and taste, like touch, tend to be categorized as exteroceptive, which refers to them being concerned with information external to the body. On this kind of distinction, proprioception (and kinesthesia) ought to count as interoceptive, and they are often referred to as such. However under the orthodox categorization of the senses it would be more correct to think of proprioception and kinesthesia as belonging to the "somatic senses" which additionally include pain, temperature, itch and vestibular balance (Kandel et al. 2000).

Interoception is distinct from the other senses which take the internal body as their object (proprioception and kinesthesia) as it is typically used to refer to the afferent sensory information from the autonomic nervous system, such as heart muscle, other smooth muscle (but not skeletal muscle which is included in the somatic nervous system), and the exocrine glands (i.e. sweat glands, saliva glands, stomach, liver, pancreas). Craig (2002, 2003a, b) however has argued against this traditional way of categorizing the senses, advocating that "interoception should be redefined as the sense of the physiological condition of the entire body, not just the viscera" (2002, p. 655). The reason for this is that recent research suggests that pain, temperature, and light touch are mediated by the same tracts in the spinal cord and subcortex as visceral information (Craig 2002, 2003b; Craig and Blomqvist 2002).

These two different ways that interoception is defined, i.e., as the sense of the visceral body and the sense of the entire physiological body, are underpinned by different mechanisms in organisms that have developed nervous systems for mediating the sensorimotor signals. After all, a cell's 'entire physiological body' just *is* its visceral body. In humans, however, this difference in the term's scope is important. While changes in the viscera that threaten the system's viability are dealt with by adapting internally, on their own they do not seem to intrinsically motivate action, at the very least not within the time scales with which we would normally judge agency. The receptors that mediate pain, light touch, and temperature, however, are hooked up so that not only can activation induce a spinal reflex (in very short time-scales) but their shared afferent pathway (with the visceral neurons) projects to motor areas in the brain such as the limbic motor cortex (ACC) for activation of movement over (still short but) longer time-scales (see Craig 2002, for a comparison of his 'spinothalamocortical pathway' with the conventional pain pathways). This interoceptive-motor loop grounds what Craig calls 'homeostatic emotions' which arise from threats to the viability of the system which homeostasis on its own cannot resolve (Craig 2003a). Craig considers these to be homeostatic 'emotions' because both motivations *and* sensations are generated. In addition to projections from this pathway to the limbic motor cortex there are also projections to the limbic sensory cortex (insula) and in primates to the interoceptive cortex and from there to the right anterior insula (Craig 2003a, b). This is of particular interest to us here because activity in the anterior insula cortex is consistently associated with subjective feelings and may even provide a basis for Sherrington's 'material me' (Craig 2003b). This supports our argument that agency and affectivity are fundamentally entwined and not only at the level of (first-order) autopoietic systems but also in complex organisms made up of various intersecting autonomous networks like ourselves – a meshwork of selfless (i.e., non-reified) selves (Varela 1991).

These insights from the neurobiology of interoception in primates allow us to gesture towards how it might be operationalized. An abstraction from its particular instantiation may be used to understand the different levels of complexity of agency in biological systems and their intrinsic affectivity. Furthermore it can guide us in modeling agency for the purposes of artificial cognitive systems research and robotics. For now we wish to emphasize one particular point: while the instantiation of an interoceptive nervous system does not necessarily give an organism *more* agency than simpler agentive organisms (at least from what we have proposed so far – either you are an agent or you are not), it does give it a qualitatively different kind of agency. This kind of agency is not merely affective but instantiates a reflective stance that is discussed by phenomenologists in terms of subjectivity. The subject's lived body is no longer always transparently lived through as the background for activity in the world, because this agential world-directed perspective can be turned on itself – inwardly – such that the body becomes an explicit part of the subject's experiential world rather than its implicit mode of revealing that world. In this way a new horizon of actions opens up, involving the development of new forms of mediated self-regulation. Thus, if we wanted to think of agency as admitting of degrees, we could emphasize this more pronounced diversity and asymmetry in the domain of agent-environment relations, which puts even more weight on the side of the body of the agent, as an enhancement of agency (Stapleton and Froese 2015).

According to the enactive theory of agency, all agents are affective systems because of the normativity involved in their adaptivity and sense-making. But it takes a special form of decoupled or mediated regulation of internal and interactive processes in order for an agent to be a cognitive agent (Barandiaran and Moreno 2006). Following what we have said about interoception, it seems plausible to suggest that such a cognitive agent must at the same time be a feeling agent, i.e. an agent that is able to make sense of the status of its internal processes. Jonas ([1966] 2001) proposed that emotions co-emerge with distal perception in order to enable the mediated regulation of actions across the greater spatiotemporal distances that perceptual experience reveals in the world, and which can no longer be bridged by mere reactive behaviour. The same logic applies to the greater internal distances revealed by interoception, whereby I perceive the state of my body at the same time as I become distanced from merely being my body, and my actions are guided by an assessment of how I feel.

It is at this point that we can start talking about subjectivity, rather than just agency, since the self, at least in a minimal embodied form, has become an explicit part of the concerns that motivate and guide actions. Much more needs to be said about this transition from the biological foundations of agency to the phenomenological conditions of subjectivity, in particular with regard to full-blown self-awareness, but we hope to have shown that enactivism is in a good position to make progress on this difficult and deep question. What is clear is that the nervous system will play a central role in this story because its electrical activity liberates an agent's capacity of regulation from underlying metabolic and developmental constraints, thereby opening up a much wider range of internal and interactive actions and perceptions, while at the same time more tightly integrating the multicellular agent as a whole (Arnellos and Moreno 2015; Barandiaran and Moreno 2006). With respect to the question of the operational requirements of specifically human subjectivity, a deeper consideration of the constitutive role of social interactions and the pre-existing cultural context will become indispensable (Stewart 2010; Kyselo 2014). Yet the biological foundations of subjectivity will remain essential even as the socio-cultural extensions of the self begin to play a significant role, which is why we are sceptical of the possibility of a genuinely collective subject – an agent consisting of a social network of other agents without material integration but having its own first-person experience (Stapleton and Froese 2015). The complex material and functional requirements for the emergence of such an integrated subjective 'self' from the perspective of a multicellular system as a whole (Arnellos and Moreno 2015) are unlikely to be repeated at the purely social level due to its looser interactions.

8.4 Conclusion

For an autopoietic system to constitute an agent there must be sufficient internal communication to allow the system to be sensitive to its own internal dynamics and move in response to them and to perturbations in the environment that threaten their boundaries of viability. Subjectivity requires something more, namely an interoceptively grounded form of sense-making. We have briefly described how this interoception is realized in the case of human agents. It is an open challenge for enactivism to operationalize this physiology in such a way that we can better grasp the essential dynamics that are being realized by the autonomic nervous system, and to determine in what manner these dynamics could be realized in other forms of embodiment. At the same time the enactive approach points beyond physiology to the essential socio-cultural dimensions of selfhood, which suggest that the emergence of human subjectivity on the basis of organismic agency cannot be fully understood in terms of changes in biological embodiment alone. The self certainly cannot be reduced to the brain alone, but neither is it limited by the boundaries of the body: there can be no self without others.

References

Arnellos, A., and A. Moreno. 2015. Multicellular agency: An organizational view. *Biology and Philosophy* 30(3): 333–357.

Barandiaran, X., E.A. Di Paolo, and M. Rohde. 2009. Defining agency: individuality, normativity, asymmetry, and spatio-temporality in action. *Adaptive Behavior* 17(5): 367–386.

Barandiaran, X., and M.D. Egbert. 2014. Norm-establishing and norm-following in autonomous agency. *Artificial Life* 20(1): 5–28.

Barandiaran, X., and A. Moreno. 2006. On what makes certain dynamical systems cognitive: A minimally cognitive organization program. *Adaptive Behavior* 14(2): 171–185.

Barandiaran, X., and A. Moreno. 2008. Adaptivity: From metabolism to behavior. *Adaptive Behavior* 16(5): 325–344.

Barbaras, R. 2002. Francisco Varela: A new idea of perception and life. *Phenomenology and the Cognitive Sciences* 1: 127–132.

Barbaras, R. 2005. *Desire and distance: Introduction to a phenomenology of perception*. Trans. P.B. Milan. Stanford: Stanford University Press.

Barbaras, R. 2010. Life and exteriority: The problem of metabolism. In *Enaction: Toward a new paradigm for cognitive science*, ed. J. Stewart, O. Gapenne, and E.A. Di Paolo. Cambridge, MA: The MIT Press.

Beer, R.D. 2000. Dynamical approaches to cognitive science. *Trends in Cognitive Sciences* 4(3): 91–99.

Bower, M., and S. Gallagher. 2013. Bodily affects as prenoetic elements in enactive perception. *Phenomenology and Mind* 4(1): 108–131.

Cameron, O.G. 2001. Interoception: The inside story—A model for psychosomatic processes. *Psychosomatic Medicine* 63(5): 697–710.

Clark, A. 1989. *Microfunctionalism: Connectionism and the scientific explanation of mental state s*. Research paper. Retrieved July 17, 2011. http://www.era.lib.ed.ac.uk/handle/1842/1332.

Clark, A. 1997. *Being there: Putting brain, body and world together again*. Cambridge, MA: MIT Press.

Clark, A. 1999. An embodied cognitive science? *Trends in Cognitive Sciences* 3(9): 345–351.

Colombetti, G. 2007. Enactive appraisal. *Phenomenology and the Cognitive Sciences* 6: 527–546.

Colombetti, G. 2010. Enaction, sense-making, and emotion. In *Enaction: Toward a new paradigm for cognitive science*, ed. J. Stewart, O. Gapenne, and E.A. Di Paolo. Cambridge, MA: The MIT Press.

Colombetti, G. 2014. *The feeling body: Affective science meets the enactive mind*. Cambridge, MA: The MIT Press.

Colombetti, G., and E. Thompson. 2008. The feeling body: Toward an enactive approach to emotion. In *Developmental perspectives on embodiment and consciousness*, ed. W.F. Overton, U. Müller, and J.L. Newman. New York: Lawrence Erlbaum.

Craig, A.D. 2002. How do you feel? Interoception: The sense of the physiological condition of the body. *Nature Reviews. Neuroscience* 3(8): 655–666.

Craig, A.D., and A. Blomqvist. 2002. Is there a specific lamina I spinothalamocortical pathway for pain and temperature sensations in primates? *The Journal of Pain* 3(2): 95–101. doi:10.1054/jpai.2002.122953.

Craig, A.D. 2003a. A new view of pain as a homeostatic emotion. *Trends in Neurosciences* 26(6): 303–307.

Craig, A.D. 2003b. Interoception: The sense of the physiological condition of the body. *Current Opinion in Neurobiology* 13(4): 500–505.

Damasio, A. 1999. *The feeling of what happens: Body and emotion in the making of consciousness*. London: Vintage.

Damasio, A. 2010. *Self comes to mind: Constructing the conscious brain*. New York: Knopf Doubleday Publishing Group.

Desmidt, T., M. Lemoine, C. Belzung, and N. Depraz. 2014. The temporal dynamic of emotional emergence. *Phenomenology and the Cognitive Sciences* 13(4): 557–578.
Di Paolo, E.A. 2005. Autopoiesis, adaptivity, teleology, agency. *Phenomenology and the Cognitive Sciences* 4(4): 429–452.
Di Paolo, E.A. 2009. Extended life. *Topoi* 28(1): 9–21.
Di Paolo, E.A. 2010. Robotics inspired in the organism. *Intellectica* 1–2(53–54): 129–162.
Di Paolo, E.A., M. Rohde, and H. De Jaegher. 2010. Horizons for the enactive mind: Values, social interaction, and play. In *Enaction: Toward a new paradigm for cognitive science*, ed. J. Stewart, O. Gapenne, and E.A. Di Paolo. Cambridge, MA: MIT Press.
Di Paolo, E., and E. Thompson. 2014. The enactive approach. In *The Routledge handbook of embodied cognition*, ed. L. Shapiro et al. New York: Routledge Press.
Egbert, M.D., X. Barandiaran, and E.A. Di Paolo. 2012. Behavioral metabolution: The adaptive and evolutionary potential of metabolism-based chemotaxis. *Artificial Life* 18: 1–25.
Froese, T. 2010. From cybernetics to second-order cybernetics: A comparative analysis of their central ideas. *Constructivist Foundations* 5(2): 75–85.
Froese, T. 2012. From adaptive behavior to human cognition: A review of *enaction. Adaptive Behavior* 20(3): 209–221.
Froese, T. 2014. Bio-machine hybrid technology: A theoretical assessment and some suggestions for improved future design. *Philosophy & Technology* 27(4): 539–590.
Froese, T. in press. Life is precious because it is precarious: Individuality, mortality, and the problem of meaning. In Representation and reality: *Humans, animals and machines*, ed. G. Dodig-Crnkovic and R. Giovagnoli. Berlin: Springer.
Froese, T., and E.A. Di Paolo. 2011. The enactive approach: Theoretical sketches from cell to society. *Pragmatics & Cognition* 19(1): 1–36.
Froese, T., and S. Gallagher. 2010. Phenomenology and artificial life: Toward a technological supplementation of phenomenological methodology. *Husserl Studies* 26(2): 83–106.
Froese, T., and J. Stewart. 2010. Life after Ashby: Ultrastability and the autopoietic foundations of biological individuality. *Cybernetics & Human Knowing* 17(4): 83–106.
Froese, T., N. Virgo, and T. Ikegami. 2014. Motility at the origin of life: Its characterization and a model. *Artificial Life* 20(1): 55–76.
Froese, T., and T. Ziemke. 2009. Enactive artificial intelligence: Investigating the systemic organization of life and mind. *Artificial Intelligence* 173(3–4): 366–500.
Fuchs, T. 2011. The brain – A mediating organ. *Journal of Consciousness Studies* 18(7–8): 196–221.
Gallagher, S. 2005. *How the body shapes the mind*. New York: Oxford University Press.
Gallagher, S. 2012. *Phenomenology*. Basingstoke: Palgrave Macmillan.
Gallagher, S. 2013. The socially extended mind. *Cognitive Systems Research* 25–26: 4–12.
Hanna, R., and E. Thompson. 2003. The mind-body-body problem. *Theoria et Historia Scientarum* 7(1): 23–42.
Heidegger, M. [1929] 1995. *The fundamental concepts of metaphysics: World, Finitude, Solitude*. Bloomington: Indiana University Press.
Husserl, E. [1952] 1989. *Ideas pertaining to a pure phenomenology and to a phenomenological philosophy. Second Book: Studies in the phenomenology of constitution* Trans. R. Rojcewicz and A. Schuwer. Dordrecht: Kluwer Academic Publishers.
Jonas, H. [1966] 2001. *The phenomenon of life: Toward a philosophical biology*. Evanston: Northwestern University Press.
Kandel, Eric R., James H. Schwartz, and Thomas M. Jessell. 2000. *Principles of neural science*, 4th ed. New York: McGraw-Hill Medical.
Kessel, T. 2011. *Phänomenologie des Lebendigen: Heideggers Kritik an den Leitbegriffen der neuzeitlichen Biologie*. Freiburg: Karl Alber.
Kyselo, M. 2014. The body social: An enactive approach to the self. *Frontiers in Psychology* 5: 986. doi:10.3389/fpsyg.2014.00986.
Lymer, J. 2011. Merleau-Ponty and the affective maternal-foetal relation. *Parrhesia* 13: 126–143.

Maturana, H.R., and F.J. Varela. 1987. *The tree of knowledge: The biological roots of human understanding*. Boston: Shambhala Publications.
McGregor, S., and N. Virgo. 2011. Life and its close relatives. In *Advances in artificial life: 10th European conference, ECAL 2009*, ed. G. Kampis, I. Karsai, and E. Szathmáry. Berlin: Springer.
Merleau-Ponty, M. [1942] 1983. *The structure of behavior*. Pittsburgh: Duquesne University Press.
Moreno, A., and A. Etxeberria. 2005. Agency in natural and artificial systems. *Artificial Life* 11: 161–175.
Panksepp, J. 1998. *Affective neuroscience: The foundations of human and animal emotions*. New York: Oxford University Press.
Parisi, D. 2004. Internal robotics. *Connection Science* 16(4): 325–338.
Plessner, H. [1928] 1975. *Die Stufen des Organischen und der Mensch: Einleitung in die philosophische Anthropologie*. Berlin: Walter de Gruyter & Co.
Reddy, V. 2003. On being the object of attention: Implications for self-other consciousness. *Trends in Cognitive Sciences* 7(9): 397–402.
Rohde, M., and J. Stewart. 2008. Ascriptional and 'genuine' autonomy. *BioSystems* 91(2): 424–433.
Sartre, J.-P. [1960] 2004. *Critique of dialectical reason. Volume One: Theory of practical ensembles*. Trans. A. Sheridan-Smith. London: Verso.
Scheler, M. [1928] 2008. *The human place in the cosmos*. Trans. M.S. Frings. Evanston: Northwestern University Press.
Sherrington, C. 1948. *The integrative action of the nervous system*. Cambridge: Cambridge University Press.
Stapleton, M.L. 2012. *Proper embodiment: The role of the body in affect and cognition*. PhD dissertation. University of Edinburgh. Retrieved from Edinburgh Research Archive: http://hdl.handle.net/1842/6396.
Stapleton, Mog, and Tom, Froese. 2015. Is collective agency a coherent idea? Considerations from the enactive theory of agency. In *Collective agency and cooperation in natural and artificial systems*, ed. Catrin Misselhorn, 219–36. Philosophical Studies Series 122. Springer International Publishing, Cham. http://link.springer.com/chapter/10.1007/978-3-319-15515-9_12.
Stewart, J. 1992. Life = cognition: The epistemological and ontological significance of artificial life. In *Toward a practice of autonomous systems: Proceedings of the first European conference on artificial life*, ed. F.J. Varela and P. Bourgine. Cambridge, MA: MIT Press.
Stewart, J. 2010. Foundational issues in enaction as a paradigm for cognitive science: From the origin of life to consciousness and writing. In *Enaction: Toward a new paradigm for cognitive science*, ed. J. Stewart, O. Gapenne, and E.A. Di Paolo. Cambridge, MA: The MIT Press.
Thompson, A. 1995. Evolving electronic robot controllers that exploit hardware resources. In *Advances in artifical life: Third European conference on artificial life*, ed. F. Morán, A. Moreno, J.J. Merelo, and P. Chacón. Berlin: Spinger.
Thompson, E. 2007. *Mind in life: Biology, phenomenology, and the sciences of mind*. Cambridge, MA: Harvard University Press.
Thompson, E. 2011. Reply to commentaries. *Journal of Consciousness Studies* 18(5–6): 176–223.
Thompson, E., and M. Stapleton. 2009. Making sense of sense-making: Reflections on enactive and extended mind theories. *Topoi* 28(1): 23–30.
Torrance, S., and T. Froese. 2011. An inter-enactive approach to agency: Participatory sense-making, dynamics, and sociality. *Humana. Mente* 15: 21–53.
Varela, F.J. 1979. *Principles of biological autonomy*. New York: Elsevier North Holland.
Varela, F.J. 1991. Organism: A meshwork of selfless selves. In *Organism and the origins of self*, ed. A.I. Tauber. Dordrecht: Kluwer Academic Publishers.
Varela, F.J. 1996. Neurophenomenology: A methodological remedy for the hard problem. *Journal of Consciousness Studies* 3(4): 330–349.

Varela, F.J. 1999. The specious present: A neurophenomenology of time consciousness. In *Naturalizing phenomenology: Issues in contemporary phenomenology and cognitive science*, ed. J. Petitot, F.J. Varela, B. Pachoud, and J.-M. Roy. Stanford: Stanford University Press.

Varela, F.J., and N. Depraz. 2005. At the source of time: Valence and the constitutional dynamics of affect. *Journal of Consciousness Studies* 12(8–10): 61–81.

Varela, F.J., H.R. Maturana, and R. Uribe. 1974. Autopoiesis: The organization of living systems, its characterization and a model. *BioSystems* 5: 187–196.

von Uexküll, J. 1909. *Umwelt und Innenwelt der Tiere*. Berlin: Julius Springer.

von Uexküll, J. [1934] 1957. A stroll through the worlds of animals and men: a picture book of invisible worlds. In *Instinctive behavior: The development of a Modern Concept*, ed. C.H. Schiller. New York: International Universities Press.

Weber, A., and F.J. Varela. 2002. Life after Kant: Natural purposes and the autopoietic foundations of biological individuality. *Phenomenology and the Cognitive Sciences* 1: 97–125.

Ward, D., and M. Stapleton. 2012. Es are good: Cognition as enacted, embodied, embedded, affective and extended. In *Consciousness in Interaction: The role of the natural and social context in shaping consciousnes*, ed. F. Paglieri. Amsterdam: John Benjamins Publishing Company.

Wheeler, M. 2005. *Reconstructing the cognitive world the next step*. Cambridge, MA: MIT Press.

Wheeler, M. 2010. Minds, things and materiality. In *The cognitive life of things: Recasting the boundaries of the mind*, ed. L. Malafouris and C. Renfrew. Cambridge: McDonald Institute for Archaeological Research.

Wheeler, M. 2011. Mind in life or life in mind? Making sense of deep continuity. *Journal of Consciousness Studies* 18(5–6): 148–168.

Zahavi, D. 1999. *Self-awareness and alterity: A phenomenological investigation*. Evanston: Northwestern University Press.

Zahavi, D. 2011. Mutual enlightenment and transcendental thought. *Journal of Consciousness Studies* 18(5–6): 169–175.

Chapter 9
Radicalizing the Phenomenology of Basic Minds with Levinas and Merleau-Ponty

Matt Bower

9.1 Radicalism, Basic Minds, and Phenomenology

Research in philosophy of mind and cognitive science carried out in the last two and a half or so decades under the banner of embodied and enactive cognition (EEC) has increased tremendously, and so it has become an important task to differentiate various currents within the larger relatively cohesive movement. While one can perhaps carve up the theoretical landscape in more than one way (i.e., with respect to distinct but relatively general problems facing researchers), one natural way of sorting things out would be to compare views within EEC concerning their conceptions of the basic ingredients that make for a cognitive process and the fundamental concepts for understanding any process insofar as it is a cognitive process. Think of this as the problem of minimal cognition (Beer 1996, 2003; Barandiaran and Moreno 2006; van Duijn, et al. 2006) or basic mentality (Hutto and Myin 2013).

Both intramural scuffles and skirmishes with those not committed to the EEC project have from early on tended in just this direction (e.g., Brooks 1991; Clark and Toribio 1994; Van Gelder 1995). And the major points of contention have been about whether cognition need be conceived in essence as information processing, understood in terms of computational algorithms applied to representational mental states, or what I'll simply call the computational-representational view of cognition (CRC).

The clashing of views about how to understand cognition in its fundaments has come to a head with the recent *radical* proposals of Chemero (2009) and Hutto and Myin (2013). They set their views apart from others working within the field of EEC by proposing that cognition be understood as radically embodied and enactive, where being radical means getting to the root of the problem and, ultimately, arguing

M. Bower (✉)
Texas State University, San Marcos, TX, USA
e-mail: membower@gmail.com

M. García-Valdecasas et al. (eds.), *Biology and Subjectivity*,
Historical-Analytical Studies on Nature, Mind and Action 2,
DOI 10.1007/978-3-319-30502-8_9

that cognition is not essentially representational or computational. Chemero deems CRC to be unenlightening for explaining cognitive phenomena, provided we can produce models of them using the resources of dynamical systems theory and can further flesh those models out with concepts drawn from J.J. Gibson's ecological psychology. Hutto and Myin entertain a more philosophical worry about CRC, claiming that the most plausible, naturalistically inclined theories of representation actually fail in their naturalistic aspirations. Lest we condemn ourselves to some unpalatable anti-naturalism, we had better jettison representations from the theory of basic mentality.

There are, it seems to me, at least three related features that make the radical EEC approach to basic minds attractive. First, the radical view removes some of the mystery that otherwise surrounds claims that cognitive processes might be body- or world-involving. This idea is difficult to conceive – though, of course, not impossible (e.g., Wilson 2004; Clark 2008) – if we think that cognition is essentially representational and computational, while the body and the mind typically operate according to some other set of principles. In short, CRC nourishes the sort of intuitions to which internalist and brain-centric critics of EEC tend to appeal. All the better for EEC to leave them behind.

Second, and by the same token, abandoning CRC when it comes to basic mentality fits better with the view that life and mind are continuous. Surely, we need categories and explanatory tools unique to cognitive phenomena. Nevertheless, for basic minds at least, we should not expect any wild leaps to arrive in the evolutionary process from life to mind. If there is deep continuity between life and mind (a là Thompson 2007; Barandiaran and Moreno 2008), then cognitive systems should, rather than being governed by a completely autonomous set of principles, be a variant and novel adaptation of the basic principles to which living systems are beholden (Di Paolo 2005).[1]

Third, and, again, in the same spirit as the preceding, a radical view of basic minds more easily lends itself to a naturalistic explanation than the alternative. One important source of trouble facing the project of naturalizing basic cognition is eliminated by going radical and rejecting the CRC view of basic minds. Given that representations as such possess semantic content – expressible in the form of predicative statements – and that the normativity of such content fails to map onto biological norms, a genuinely "hard problem" arises for those who hold that cognition is fundamentally representational (Hutto and Myin 2013). No such problem exists for the radical strand of EEC.

Given the productive exchanges and intellectual cross-pollination that has taken place in recent years between phenomenology and the philosophy and sciences of the mind, I would like to explore here whether phenomenology may be able to lend a helping hand to the radicals. It actually can, I hope to show, by giving us a clearer picture of basic minds from the person-level and in light of the lived-experience of subjects equipped with (among other things) basic minds. The following reflections

[1] Sure enough, radical enactivists like Hutto have a favorable attitude toward their autopoietic enactivist cousins (2011).

on phenomenology and basic minds will contain two primary emphases. It will center, negatively, on the issue of representation (or the lack thereof) in basic minds and, positively, on their characteristically embodied, non-representational forms of constitutively world-involving intentionality.

Note, I will leave aside the issue of computation.[2] The computational theory of mind understands computational processes to take place at the subpersonal level. Although I will make no proposals, it may be that phenomenology has something to contribute here. Nevertheless, phenomenology's primary subject matter is person-level cognitive processes. Experiences involving conscious awareness form the core of this subject matter, though phenomenology is not restricted to them, inasmuch as, for some phenomenologists at least (e.g., Heidegger, Merleau-Ponty, Levinas), the careful study of consciously aware states can be revelatory of broader intentional dynamics that don't involve conscious awareness. Our concern here, then, will be with whether person-level features of cognition are best understood in representational terms, and, if not, how else they might be understood.

The resources of phenomenology available to proponents of the radical EEC view of basic minds are vast. Many working within the phenomenological tradition in twentieth century European philosophy have devoted themselves to giving a theory of basic mentality. This is true even of Franz Brentano, a major influence on Husserl's phenomenology, and it is one of the reasons he is still discussed today. He hoped to provide a distinct conceptual-explanatory framework for psychology, and proposed intentionality as the basic characteristic of cognitive phenomena as such. The same concern animates Husserl from his earlier work in the *Logical Investigations*, which contains critical but largely favorable discussions of Brentano's view on the matter, and his last work, the Crisis, in which he continues to maintain with Brentano – but in his own way, of course – that psychology's fundamental concept is that of intentionality.

Although post-Husserlian phenomenology is overall much less concerned with the issue of the fundamental categories of psychology and how to understand its subject matter than Husserl was, the dialectic within the phenomenological movement tends to return again and again to the theme of intentionality, whether in a critical spirit or otherwise. Heidegger, for instance, claims that Husserl's theory of intentionality is insufficient, and rests on a more basic and hitherto overlooked phenomenon, namely, that of "transcendence" (Heidegger [1927] 1962; Moran 2014). Merleau-Ponty ([1964] 2003) makes a similar move in introducing notions like "body schema" and "intentional arc," though, unlike Heidegger, Merleau-Ponty shares Husserl's interest in engaging the sciences of the mind. And, again, Levinas and Henry both make their mark on the phenomenological landscape in no small part thanks to their appreciative but deeply critical reflections concerning the Husserlian theory of intentionality (Levinas [1961] 1969; Henry [1990] 2008).

[2] Even some who deny that representational explanations are necessary for certain basic cognitive processes (e.g., Orlandi 2014; Anderson 2014) allow that computation of some (non-classical, i.e., non-algorithmic) sort may nevertheless be involved in subpersonal cognitive processing.

Given this abundance of pertinent material, the brief account offered here will be highly selective. I will first review, in the next section, some ideas from Husserl and Heidegger. Their understanding of the basic character of our directedness toward the world, while often appearing amenable to the radical EEC view, lends itself to being interpreted in representationalist terms. This serves as a reminder that the phenomenological movement is rife with internal disagreements. Phenomenological errors, though, are open to phenomenological critique and correction. The discussion of Husserl and Heidegger will thus be a foil for the remainder of what follows, where I will offer some more constructive suggestions about how phenomenology might lend support to the radical EEC approach to basic minds.

9.2 False Starts: Representation in Husserl and Heidegger

In this section, I will review somewhat schematically the accounts of our basic form of world-directedness[3] formulated in the phenomenologies of Husserl and Heidegger. In so doing, I will explain why their accounts commit them to a representationalist view of basic mentality. In both cases, even if they offer up diverging or even conflicting stories about our basic world-directedness, I hope to show that they both nevertheless agree that our basic involvements with the world have semantic content, which makes them truth-conditional and, hence, representational.

Husserl's phenomenology consists in large part of a theory of intentionality. The ground level of this theory is an analysis of the intentionality embedded in perceptual experience. For Husserl, what needs to be explained to make sense of perceptual intentionality is how we pick out individual items in the flux of sensory experience. The function of perception is thus to identify objects, and a particular episode of perception lasts just as long as one remains directed toward some selfsame entity. Hence, he conceives of perception (along, ultimately) with all other forms of intentionality), insofar as it is inevitably a temporally thick happening, as a synthesis of identification (1999, pp. 79–80, 41–42).

A perceptual synthesis of identity has two sides or "poles" to it (1999, pp. 77–79, 39–41, 1983, §§97–92). On one side, there are those characteristics that distinguish perceptual directedness from other forms of intentional directedness to the same object, e.g., in memory or abstract thought. These are dubbed noetic. The way an object – a die, to use Husserl's example – appears in perception differs from how it might appear in remembering that very same object or discussing it prospectively in its absence with someone else. On the other side, there are those characteristics of the object through which it is presented or might be presented, by virtue of which we are able to single it out and maintain a synthesis of identification over time. These are called noematic, and it is these that we need to consider more closely at present.

[3] I will use the phrase world-directedness instead of instead of intentional directedness due to the reticence of Heidegger and Levinas about the notion of intentionality.

A momentary percept will contain an immensely rich presentation of the perceived object. The perceiving act picks out of this concrete array relatively indeterminate but determinable characteristics. For instance, when presented with a die lying on a table, three of its sides presented, with a particular color, illuminated just so, and bearing certain markings on its sides. Now, when the perceiver moves around the die or picks it up to reveal other sides, all of these concrete characteristics will change. But there will be a core of continuity throughout that change, a continuity consisting of relatively indeterminate characteristics that the object possesses regardless of the perceiver's momentary point of view, e.g., cube-shaped, rigid, colored, containing dots on its sides, etc. The stable presence of these more generic characteristics – the overall "agreement of genus," as Drummond says (1990, p. 155) – is supposed to enable a synthesis of identification across the instable flux of concrete sensory experience.

Tellingly, the collection of all possible noematic characteristics for a given object is termed the object's noematic sense, and Husserl often casually refers to an individual noematic characteristic as a sense. As to perception, he states: "Perception, for example, has its noema, most basically its perceptual sense, i.e., the perceived as perceived" (1982, pp. 182, 214). Clarifying what such a sense might contain, he continues: "'In' the [...] perception [...], we find, as indefeasibly belonging to its essence, the perceived as perceived, to be expressed as 'material thing,' 'plant,' 'tree,' 'blossoming'; and so forth" (1982, pp. 184, 216). There is much debate surrounding precisely how to understand the noema,[4] but this much seems clear: Intentional directedness to objects in perception is secured by means of perceptual sense, which is a relatively abstract description or set of descriptions of the object. In short, perception targets its objects under certain modes of presentation or descriptions.

However these descriptions might be understood – e.g., whether they are some sort of curious ideal or intensional entity (Smith and MacIntyre 1984), or are, in some equally curious manner, properties belonging to the perceived object (Drummond 1990) – it seems to me that their involvement in perceptual intentionality renders the latter representational. If the noema turns out to be an intensional entity, then surely it is representational, and if not, the idea that semantic content "ain't all in the head" is now a familiar one (Putnam 1975; Burge 1979), so embedding the noema in (or identifying them with) things is at least consistent with its being representational (Smith 2008).[5] That is because they burden perceptual intentionality with semantic content: Perceiving tracks the "perceived *as* perceived," the targeted sensory configuration *as* a tree, the tree *as* blossoming, etc. The "as" inherent in perception gives it semantic content to the extent that being so structured opens up the possibility of truth or falsehood, i.e., to the extent that the as-structure is understood to be truth conditional. To perceive an object, then, is to *represent* it

[4] See Drummond and Embree (1992) for a variety of viewpoints on the issue.

[5] Though, that likely doesn't entail for Husserl a wholesale commitment to the entire suite of vices characteristic classic GOFAI-style representationalism. See Yoshimi (2009).

as being thus and so, and for the further course of experience to be a matter of confirming or informing that supposition.

There may be a great deal, at least on the surface, that separates the phenomenology of Heidegger from that of Husserl, but they share a common commitment to the idea that our basic form of world-directedness is laden with an "as structure." Heidegger's account differs from Husserl's in part due to Heidegger's de-emphasis of the significance of particular acts (e.g., of perception), and corresponding valorization of more basic tendencies and dispositions that lie at the basis of particular engagements with our surroundings. One such basic disposition is what Heidegger terms "understanding" (1962, §31). A being, Dasein (which ends up being coextensive with humanity), is characterized as understanding just in case it has a pre-reflective grasp of its potentiality for being. Dasein doesn't antedate the possible ways of getting along that constitute this potentiality for being, but always finds itself "projected" onto them. That is, roughly, a key aspect of the basic human predicament is the ineluctable and ineliminable drive to forge a practical identity, trying to be a person of a certain sort (Crowell 2013).

Understanding, then, is not directed immediately or primarily towards one's surroundings and the objects ensconced therein, but toward oneself. Nevertheless, being projected onto one's possibilities opens up a world for Dasein. One's projected possibilities determine the significance that any particular encounter with one's surroundings might take on. The meaning of what happens to be on offer for one in the world turns on the sort of person one happens to be and is seeking to be. Dasein's understanding disposition thus first determines the character of its overall situation, which is the necessary pragmatic background against which particular worldly exchanges take place. The generic, default characteristic of worldly items is for them to be "ready-to-hand," i.e., to play a functional role along with a network of other items in some possible task (Heidegger 1962, §15).

The basic form of world-involvement that emerges in Heidegger's story is that of circumspective navigation of a situation, an opportunity for action rife with equipment suitable to be taken up in that action. As puts it, "the less we just stare at [e.g.] the hammer-Thing, and the more we seize hold of it, the more primordial does our relation to it become, and the more unveiledly is it encountered as that which it is – as equipment" (1962, pp. 98, 69). Dasein's understanding disposition, then, ultimately sets the stage for particular encounters. Understanding "develops itself," i.e., enacts certain of its possibilities, in this or that circumstance in the form of a pre-predicative "interpretation" (1962, §32). Interpretation makes something understood in terms of one's extant understanding of oneself and one's situation.

Along with the notion of interpretation, Heidegger's account takes on a semantic character very similar to what we saw in Husserl above. Consider Heidegger's remark that "[w]hen entities [...] have come to be understood [...] we say that they have meaning [*Sinn*]. But that which is understood, taken strictly, is not the meaning but the entity [...]. Meaning is that wherein the intelligibility of something maintains itself" (1962, pp. 151, 192–193). Moreover, he states that the way an entity shows up in interpretation "has the structure of something as something" (1962, pp. 149, 189).

The idea seems to be this: Dasein's understanding endows a given scene with the character of a situation, a place where something is to be done equipped with an array of items to be navigated or employed to that end; Dasein's interpretive regard picks out items within or features of that situation *as such* by attending to their functional properties (e.g., "with which," "in which," "in-order-to", etc.). The latter are the descriptions that constitute the content of Dasein's circumspective interpretation (see Wrathall 2011, pp. 52–55). Despite the distaste of this consequence for some of Heidegger's interpreters (Carman 2003; Wrathall 2011), Heidegger gives us no reason to doubt that this is *semantic* content, expressible in propositional form and truth conditional in nature (Heidegger 1995, pp. 415–418, 287–288; Golob 2013). So, Dasein's understanding may not be representational, but when the rubber meets the road in Dasein's real-time worldly interactions, its interpretations make a claim on the world as being thus and so. It represents entities in functional terms, e.g., the hammer as grasp-able, for-hammering, etc. And, to emphasize the pervasive presence of interpretation in Dasein's experience, Heidegger claims, were one to encounter an item less the interpretive "as," that would count precisely as "a failure to understand it any more" (1962, pp. 149, 190).[6]

9.3 Levinas' Sensibility: Embodied Intentionality Without Semantic Content

Now that we have seen why Husserl and Heidegger conceive of basic mentality in representational terms, I want to present an alternative conception appearing in the work of Levinas. Levinas, admittedly, does not deny that we *can* enjoy perceptual experiences or circumspective interpretations of our surroundings more or less as Husserl and Heidegger describe them. There is, in fact, a place for representations even in the radical EEC picture of things – just not within the domain of basic mentality. Levinas, too, denies that these sorts of representational episodes, laden as they are with semantic content, are fundamental or that the representational elements they countenance are essential to our basic world-directedness (Levinas 1998, p. 165; Morgan 2011, p. 88). He makes his case by illustrating descriptively the limits of those frameworks, exposing a domain of experience that operates in a fundamentally different manner.

Levinas' most well-known criticism of representation comes as part of his theory of interpersonal understanding in second-personal contexts. While this issue won't be our main focus, it's worth mentioning, because the radical EEC view of basic

[6] Despite these critical remarks, the phenomenologies of Husserl and Heidegger are still full of valuable insights, even for the radical EEC project. See, e.g., Dotov et al. (2010). I think non-trivial amendments are needed to radicalize their theories of perception and understanding/interpretation, but those are amendments worth making. Elsewhere, I have made this case with respect to Husserl's phenomenology of perception (Bower 2014). I also think Heidegger's anti-representationalist interpreters have contributed helpfully to the same end.

minds definitely ought to and in fact does extend to include an account of social cognition (Hutto 2008; Gallagher 2008; Gallagher and Hutto 2008), and one that has affinities with Levinas' ideas (Gallagher 2014). Levinas' main philosophical interest lies in the issue of how individuals relate to one another face to face, a relation Levinas believes to be fundamentally ethical. There are several features of representation that Levinas maintains are inadequate to this task (MacAvoy 2005). Consider just two, sketched in very broad strokes. On the one hand, the meaning one's representation attributes to another always seems to miss its enigmatic target. The face to face encounter with another person has a tendency to reveal the limitations of our representational capacities in the "resistance to the grasp" that others present to us (Levinas 1969, p. 197). They defy the descriptions we use in fixing reference to them, since *in the face to face encounter* they simply do "not [fit] under a category" at all (1969, p. 69).

Second, our very ability to represent at all in the characteristic way that we do – by means of a whole system of meanings (i.e., in some natural language) that we freely use to exchange information with others – must itself rest on a more basic second-personal stance or "orientation" toward others (1987, pp. 75–108). To share meanings with others and communicate in a common representational medium takes for granted an ethical attitude, a desire to make peace and come to terms with them rather than resorting to violence. And, as the condition for our ability to represent at all, this ethical orientation must be non-representational in character.

Our concern, though, is not with interpersonal understanding and ethics, but with other core aspects of our basic form of world-directedness. Although it is a much less prominent theme in the literature, Levinas does go to considerable lengths to depict for us more generally what a life without representation might be like. It is not only the face-to-face encounter that defies a representation-riddled explanation, but just as much our basic mode of world-directedness, which he often calls sensibility. Sensibility, "which is the very consciousness of a living being," he baldly states, "is not a thought which is confused; it is not a thought at all" (1987, pp. 26, 127). And of the world as engaged in sensibility he remarks: "We live from 'good soup,' air, light, spectacles, work, ideas, sleep, etc... These are not objects of representations" (1969, p. 110). Perhaps, one will say, sensibility so construed is not too far from a Heideggerian, pragmatic view of representation in our basic world-directedness. Levinas, however, avers that the categories of practical reason are no more applicable here than those of speculatively or theoretically driven reason (1969, p. 113). There is no doubt that, at least from his own point of view, Levinas' view on this point does not map on to Heidegger's (Hand 2009, p. 41).

It is worth being specific about what aspects of representation Levinas deems inappropriate to attribute to sensibility. Representation is, on the one hand, a unilateral affair. It is driven by the activity of the subject, being the product of the subject's "mastery" (1969, p. 124). Unilateral control is a crucial feature of representation to the extent that representation is an anticipatory or expectation-guided process. The continuity of noematic sense in an ongoing perceptual experience, for Husserl, is guaranteed by future-directed "protentions" partly constituting its intentionality (1999, §19). Heidegger, similarly, maintains that the meaningfulness of entities we meet circumspectively in our surroundings is contained in advance of particular encounters due to the "fore structure" of the understanding (1962, pp. 150–151, 191–192).

A consequence of this is that "[t]he act of representation discovers, properly speaking, nothing before itself" (Levinas 1969, p. 125, 1998, pp. 125, 161). Whatever we perceive or cirumspectively encounter is only presented under some description, i.e., the semantic content of our momentary world-directedness (1998, p. 128). This content is possessed prior to the particular instances of its world-involving deployment, whether as a familiar type (Husserl 1973, §§8–9) or as a structural wrinkle in out totality of involvements with things (Heidegger 1962, pp. 149–189). It can be presented with a greater, unanticipated determinacy in novel circumstances, but the basic character of the experience is one of recognition or re-identification, if only by type.

By contrast, "sensibility touches the reverse, without wondering about the obverse" (Levinas 1969, pp. 135, 131; Vasseleu 1998, pp. 80–81). Put somewhat differently, sensibility is not expectation-guided. In sensibility one enjoys the full concreteness of the sensory array and relates to it without isolating its more abstract properties or being carried off (at least tacitly) into one's projected possibilities. A characteristic modality of sensibility is enjoyment, gratifying one's needs by assimilating one's surroundings. Of course, we do so in obvious ways when we draw in air or enjoy a meal. It might seem, then, that enjoyment is derivative of a more primitive activity that puts us in a position to enjoy things. We represent, then we enjoy the things we represent. One can, the thought is, only enjoy what one already recognizes in one's presence or seeks out. Levinas holds, however, that even sensory access to our surroundings should be understood as a form of enjoyment (1969, p. 130, 1991, pp. 72–73). Vision is just as need-driven as the desire for food; in vision, one "leaps over" the play of expectations in order to devour a scene with one's sight.

Sensibility, additionally, is not best understood as an exclusively unilateral exercise of control. To the contrary, it is characterized by a significant degree of "passivity" (1998, p. 130). Need impels us toward enjoyment, but, by the same token, need makes us vulnerable and indigent (1991, p. 74). The elements that nourish us are the same ones to which we may be exposed and by which we may be "put out." Whereas representation is one sided or non-reciprocal, sensibility is a function of how our surroundings meet our needs or threaten us in our fundamentally vulnerable condition. That is, sensibility is determined just as much by one's own condition as by the external milieu in which one happens to be immersed. Levinas thus describes the subject of sensibility as "autochthonous," referring to the fact that it is "enrooted in what it is not," even if "it is, nevertheless, within this enrootedness, independent and separated" (1969, p. 143). Further, enjoyment not only exceeds unilateral control and anticipation, in some instances, at least, it takes one entirely outside of one's projected practical possibilities, since, according to Levinas, "[t]o live is to play, despite the finality and tension of the instinct to live" (1991, p. 134).

Let's consider more closely what Levinas proposes to fill the explanatory void of representation and positively explain the possibility of a basic world-directedness that is not characterized by unilateral control and the guidance of expectation. We don't need to control or projectively envisage an environment that we can "assume" instead, relying on it as just the sort of place environment in which "[w]ell-trampled places do not resist me but support me" (1969, p. 138). Assuming in, e.g., visual experience means participating in a process of "vision already and henceforth borne by the very image that I see" (1969, p. 128). The key presupposition here is that sensibility is determined by one's bodily condition. First, what stands out within one's surroundings in sensibility always answers to an embodied tendency, e.g., hunger, or, perhaps, visual and haptic exploration. Second, one's surroundings are taken for granted as a niche, a site that supports the activities in which one's tendencies typically issue by putting one's body to work.

The two points warrant explication. How, after all, can we pick out in sensibility what our surroundings have on offer for us if not under some description? Sensibility is supposed to be a form of intentional directedness and not a play of mechanical forces. What is required here, if Levinas' account is correct, is a story of intentionality without representational content. This is, as it happens, the exact move Hutto (2008) is also led to make in framing his radical EEC conception of intentionality in basic minds. The key is to think of intentionality as the exercise of a tendency to selectively respond to certain items in or features of one's surroundings. The tendency is to respond to a determinate set of items or environmental features. It enables one to discriminate, and thus has a more or less determinate *extension*. But it lacks intensional, semantic content.

Viewed from the outside, this sort of intentional directedness will appear no different from picking items or features out under some description. If the sight or smell of food evokes hunger, it will be tempting to think of sensibility's response as a response to the item *as* (presented under the aspect of), e.g., being delicious-to-eat. And that's even a correct description – from the third-person perspective or in the clear light of hindsight. This temptation to take up an "intensional stance" is exactly what leads Husserl and Heidegger astray. The intentional directedness, though, only picks out the relevant item or feature, and not the semantic complex "such-and-such *as* such-and-such." I suggest this is the best way to understand Levinas' idea that sensibility takes things just as they are, in the full immediacy and richness of their sensory presentation. In a somewhat different idiom, he repeatedly refers to the worldly offerings presented in sensibility as "pure qualities without support," "adjectives without a substantive," and "content without form" (e.g. 1969, pp. 130–134, 161). To stick with the preceding example, when goaded by the body's hungry condition, sensibility fixates simply on the delicious-thing-to-eat.

Back, now, to the second point. Given that, as we just saw, sensibility allows us to discriminate and respond selectively, how might it work that, e.g., in vision sight is able to devour a scene, take it in, by "assuming"? We have to address the mutualistic character of sensibility and explain how one's environment gives direction to the enjoyment of sensibility. In the enjoyment of sensibility, "the distinction between activity and passivity is undone in agreeableness" (1969, p. 163). Here, the body is a "cross-roads of physical forces" that either threaten or nourish it, so sensibility's assuming is a matter of taking for granted that it may find a world agreeable to it, since in sensibility "[w]hat is necessary to my existence in order to subsist is [also what is] of interest to my existence" (1969, p. 164). The indigent body that enacts sensibility assumes a world fit for it, in a kind of "simultaneity [that] constitutes the body" (1969, p. 165). Representation only becomes necessary when the assumption of this fit becomes precarious, or, as is often the case with human beings, at least, when it is surpassed in order to produce a greater degree of stability and independence from environmental contingencies (1969, pp. 158–162).

Again, Levinas' conception of our basic form of world-directedness resonates with the radical EEC view. As to this second point about mutuality, what he describes seems to me to be the complementary person-level dramatization of just the themes from J.J. Gibson's ecological approach to psychology that has been so influential to proponents of radical EEC (Chemero 2009). The ecological view of perception understands it to be essentially world-involving, an interaction that directly relates one to the one's surroundings. Perception so described is non-representational, and this is because the environment fits the perceiver's needs. First, it fits the perceiver's needs because it contains just what the perceiver wants to know about things, exactly what interests it, with no need for interposing expectations or projective interpretations. It presents opportunities for action or "affordances" (Gibson 1979; Chemero 2003), e.g., delicious-to-eat.

Second, the rich sensory array by itself suffices to present these affordances. The so-called outfielder problem illustrates this idea (Wilson and Golonka 2013). In baseball, when an outfielder is tasked with catching a fly ball, the best strategy, and the one outfielders seem to actually use, is not to calculate or think in any way about the ball's starting point, trajectory, its gradual acceleration and deceleration, etc., in order to then head straight to the ball's destination. Instead, the thing to do is to keep continuously fixated on the ball as it flies and let the dynamic structure optical flow provide direction – and not in a straight line, it turns out – toward the catching site. As Wilson and Golonka explain: "The affordance property 'catchableness' is therefore continuously and directly specified by the visual information, with no internal simulation or prediction required" (Wilson and Golonka 2013, p. 6). So, to grossly oversimplify things, maintaining that contact while moving is sufficient to guide the outfielder to an appropriate site for catching the ball. This is, reiterating Levinas' phrase, an instance of vision "being borne by the very image I see" (1969, p. 128). On the radical EEC view of perception, rather than going beyond the sensory array, one leaps further in to generate the right sensory inputs (Barrett 2011) – "'enjoying a spectacle,' or 'eating up with one's eyes'" (Levinas 1991, p. 67) – assuming the world will support that activity.

9.4 Merleau-Ponty's Dynamic, Synergetic View of Basic Minds

In the phenomenology of basic mentality developed so far, we have seen in Levinas' work a way to understand our basic world-directedness in non-representational terms. Roughly, representations can be dispensed with if we are equipped with a sensibility that disposes us to respond appropriately to entities or features in our surroundings in light of our bodily needs, and if this process can be executed by relying on or assuming the world's agreeableness to our activities and needs.

In this section I want to supplement the Levinasian phenomenology with some ideas from Merleau-Ponty's phenomenology. His account of basic mentality as a "synergic system" (2003, p. 272) and of (non-semantic) sense add important details and qualifications to the mutualistic picture of basic mentality we saw in Levinas. Dwelling on the details of the synergetic view of basic mentality will both reinforce these Levinasian themes and, thus, the inadequacy of a unilateral, expectation-based representational conception of basic minds, but will also make more prominent the part of the body in the synergetic system. I will suggest, finally, that this synergetic view of basic minds is consonant with a central tenet of radical EEC views of basic mentality, namely, that minds are at bottom dynamical systems (Van Gelder 1995; Thelen and Smith 1994; Chemero 2009). As such, basic minds are open, self-organizing systems. The coherence and order exhibited by mentality emerge from its past and present interactions, and, further, the mutualism is so thoroughgoing that mind, body, and world may even be viewed as a single system.

First off, Merleau-Ponty intends his analysis of sense (and its "institution") to take the place of what Husserl called constitution, the basic schema for intentionality in all its forms. We considered an instance of constitution above while reviewing Husserl's understanding of perceptual intentionality. Husserl deals with the latter in a "form-content analysis." On the one hand, there is a flowing, array of sensory experience, and, on the other, there is the discernment of form within that array. The passing die-percept presents an infinite wealth of particular die-features (content), and the task of perception is to identify and track the stable, generic features in order to keep a perceptual hold of the die (form).

Rejecting this analysis, Merleau-Ponty states: "When we approach an object [...], there are no numerically distinct *Abschattungen* [profiles, i.e., sensory appearances] and no *Auffasung als* [apprehension as] ... representation of one selfsame intelligible core" (2010, p. 5, 2003, p. 176, n. 9). Nor is it any improvement to appeal to the Heideggerian framework. To whoever "will say it is through relation to a project" that, e.g., one encounters an object as such in perception, Merleau-Ponty responds: "If you like, but there is a non-decisionary project, not chosen, [an] intention without subject: living" (2010, p. 6). He is willing to speak with Heidegger of sense in terms of "transcendence," but the particular worldly encounters transcendence enables are "not a recognition of the concept," because sense "means a power to break forth, a productivity" (1968, pp. 209, 208).

The rejection of representation, to echo themes we've seen in Levinas, follows from the recognition of the ambiguous "passivity-activity" of our basic world-directedness. Our basic task is not to assign meanings to things, and, in any event, what is "instituted makes sense without me" (2010, p. 8). The idea of sense in this context is crucial. Whether the perceiver is the sole impetus for it or not, in perception things have sense for us. Sense of this sort is manifestly not equivalent to an intensional entity or semantic content. While those connotations naturally spring to mind, the French word *sens* used by Merleau-Ponty has other prominent connotations. *Sens* also connotes significance, direction or orientation, tendency and it is no doubt these semantic contours that make the word apt for his use of it (Morris 2004, p. 24; Marratto 2012, p. 55). Indeed, highlighting the spatial metaphor, he says that "institution [means] the establishment in an experience […] of [senses as] dimensions […] in relation to which a whole series of other experiences make sense" (2010, pp. 8–9).

Negatively, Merleau-Ponty opposes the representational view because "it cannot be said that the one [i.e., the perceiver] acts while the other [i.e., the perceived object] suffers the action, or that one confers significance on the other" (2003, p. 248). The perceiver does not have all the resources to make sense of the perceived, and the perceived tends to appear in the first instance as "a kind of muddled problem for my body to solve" (2003, p. 249). Note, though, that the body's "solution" is engendered in the interaction (Morris 2004, p. 16), and, moreover, that what Merleau-Ponty has in mind in referring to the body is the "prepersonal" life of the perceiving organism (e.g. 2003, pp. 250–251). The relation of "assuming" the world that Levinas describes reappears within the internal economy of basic minds, since the domain of conscious experience "takes for granted" and "takes advantage of work already done" by the prepersonal body (2003, p. 277). This prepersonal life will include the organism's nervous systems, and, of course, its schematic, practical sensorimotor know-how, but surely also the subpersonal monitoring of its vital condition and even the biomechanical properties of its musculoskeletal system – in short, all those elements that serve to condition, guide, shape and support action without the need for conscious awareness.

The complexity Merleau-Ponty observes at work in perception is truly impressive. Not only must one's conscious awareness be in order, but the world must do its part together with a variety of pre-personal psycho-biological systems. Furthermore, as he emphasizes especially in his later works, the body as perceived – not just proprioceptively and kinesthetically, but in touch and vision – is also a significant element of this organizational complex (1968). And the organization of these disparate ingredients takes shape without a central governing mechanism for overall control, or any common informational-representational medium of information that might be exchanged between them. Even within the domain of conscious experience, the perception of an item through multiple sensory channels or modalities displays this complexity and decentralized organization of multiple elements.

For instance, Merleau-Ponty is emphatic that tactile experiences of a spatial array "do not and never will stand in a relationship of synonymity with those of visual space" (2003, pp. 259, 261). They do "communicate" with and even "interpret" one another, but "without any need of an interpreter" (2003, p. 273), i.e., not by means of a common amodal medium. And the latter is not even something imposed downstream after sampling an item with both modalities on some occasion, because "the constitution of an intersensory world must be effected in the domain of sense itself" (2003, p. 261). How can that be? Disparate sensory modalities "communicate" by participating in a common project. It is their mobilization in action that links them together. In perceptual exploration "[t]he senses intercommunicate by opening onto the structure of the thing" (2003, p. 266). The link between any sensory channel and the others is not written in its "anatomical layout," but emerges from "the use which the psychosomatic subject makes of it" (2003, p. 268). It is the dynamics of continuous sensory sampling through multiple channels tied to task-appropriate bodily movements – themselves sensed through proprioceptive modalities – that achieves intersensory organization (1968, p. 134).

With the overall framework of Merleau-Ponty's view sketched out, let's return to the notion of (non-semantic) sense. As David Morris explains, sense "is neither a meaning in the head nor is it interior to subjectivity, it is a meaning within a movement that crosses body and world" (2004, p. 24). Sense belongs to a *situation*, where the embodied perceiver's action-tendencies and practical know-how constitute a tacit "decision" that makes possible its adaptive sensitivity to action-relevant items or features of its environment, which count as "motives" for its response (Merleau-Ponty 2003, pp. 301–302). If that is so, it is obvious that Merleau-Ponty's notion of sense bears no great resemblance to sense as a semantic notion (e.g., denoting an intensional entity). Merleau-Pontian sense is a dynamic network of relations and interrelations between elements scattered across mind, brain, and environment. Unlike the semantic content of a representation, it does not belong to any one element or have any one element as its sole cause, and it doesn't antedate the elements (Di Paolo 2015), although the prior history of the system and its elements no doubt partially determine its present behavior.

Even if it is admitted that representational vehicles may also be scattered across body and environment (Wilson 2004; Clark 2008), they nevertheless belong to the mind, the entity that owns representations (see, e.g., Rowlands 2010). Sense, on the other hand, does not belong to the mind. It cuts across the whole dynamic network of elements that basic minds are caught up in. Chemero's analysis of the Gibsonian notion of affordance, which I take to be nearly synonymous with sense, supports this point. As he understands it, perceiving an affordance amounts to "perceiving that the situation as a whole has a certain feature, that the situation as a whole supports (perhaps demands) a certain kind of action," and, in addition, that "affordances belong to animal-environment systems" (2009, p. 140, 2001).

If it turns out that those are extrinsic characteristics of representations, sense may be dissociated from representation in functional terms as well. The function of a representation is to identify distal causes of experience given the "poverty of the stimulus" in sensory experience, or to stand in for items or features of the environment that are of interest but (at least in part) presently not available for sensory sampling. What exactly is the function of sense, then? Sense is a matter of dynamic coordination and control. The perceiver relies on the native competence of its body and the stable support of its cognitive niche. The organization of the elements of the former enable the perceiver to respond to action-salient features to which the latter are sensitive. Those features serve as guides to behavioral goals as long as the perceiver can exploit its prepersonal bodily repertoire to continuously sample them or maintain the initial direction provided by those action-salient features when they momentarily disappear from view.

As I mentioned at the outset of this section, and as Di Paolo (2015) and Buhrmann and colleagues (2014) have noted, Merleau-Ponty's account of basic mentality, which we saw illustrated in his analysis of perception, strikes a chord with the radical EEC view that minds are basically dynamical systems (Chemero 2009). Radically inclined dynamical systems views of cognition, too, emphasize that it is a matter of coordination and control in the service of adaptive behavior (Anderson 2014, pp. 182–185). Of course, the simplest and most obvious resemblance is the emphasis on dynamics. In Merleau-Ponty's terms, "subjectivity, at the level of perception, is nothing but temporality" (2003, p. 278). Generally speaking, though, any system whose variables change continuously over time can, for that stretch of time, be described as a dynamical system using the formal mathematical modeling tools of dynamical systems theory (DST). A wave, a flame, even a digital computer, is a dynamical system.

Things get more interesting once we recognize that multiple dynamical systems can interact, and when they do so can be described as a single dynamical system. In that case, the two systems are coupled. The radical appropriation of DST conceives of mind/brain, body, and world as nested coupled systems (Dotov 2014). What we've seen of Merleau-Ponty's view of basic minds should already make it apparent that this suggestion is not foreign to Merleau-Ponty. In fact, according to him, body and world "form a couple, a couple more real than either of them" (1968, p. 400). In truth, if Merleau-Ponty is right, the coupling involved in basic mentality contains many interacting systems at both the personal and prepersonal levels. This coupling explains the mind's ability to coordinate with and enjoy some degree of behavior-guiding control over action-relevant environmental features. Two systems are coupled when variables of one constrain the tendencies, the attractor landscape, of the other, and *vice versa*. A relation such as that holds true on Merleau-Ponty's view of the matter, too, as can be seen when he describes the reliance of conscious awareness on the variety of tendencies comprising its prepersonal life, and, again, in the relation between that prepersonal life and the array of salient features of a given situation motivating its responses.

To illustrate, consider Esther Thelen and Linda Smith's dynamical explanation of how bodily movement and multi-channel sensory experience coordinate in perception. Their account matches closely Merleau-Ponty's analysis as described above. As they understand it, developing powers for perceptual discrimination requires both the use of multiple sensory modalities to sample one's surroundings (e.g., using both visual and tactile experience) and their time-locked coordination with kinesthetic information. "[Bodily] movement must itself be considered a perceptual system," they urge (1994, p. 193). Acquiring a perceptual grip of one's surroundings requires the virtues peculiar to both sensory and motor processes. The virtue of the former is more analytical, in sensory sampling that functions to detect relatively fine-grained features of the distal environment. When mapped onto concurrent motor processes, the detail captured by sensory processes takes on something like *Gestalt* properties. That global coherence is due to the relatively coarse-grained character of gross bodily movements captured in the flow of kinesthetic experience. Thelen and Smith cite the computer simulation devised by Reeke and Edelman (1984) to recognize alphabetic letters as an illustration of this type of sensorimotor integration that enables learning. Reeke and Edelman's device learned to pick out, e.g., a variety of permutations of the letter A without any task-specific instruction, "just from looking at and acting on specific instances" (Thelen and Smith 1994, p. 168). The dynamic interplay of sensory and motor processes was sufficient to produce this result.

Just as in Merleau-Ponty's account, the coupling explains coordination and action-guiding control without any particular system taking the lead as a central control mechanism. The perceiver's control is due in considerable part to its reliance on factors beyond its control, which it can depend on given that it has settled in a cognitive niche suitable for it. Further, there is no exchange of information, no common currency among the various coupled systems. On both Merleau-Ponty's and Thelen and Smith's understanding, integrating diverse sensory modalities need not involve a higher-order cognitive mechanism for translating each into the other's terms. Task-driven interaction gives them their coherence, and it is the overall dynamics of the entire system that make the relevant difference for both accounts.

9.5 Conclusion: Radicalizing the Phenomenology of Basic Minds

I hope to have shown that the phenomenologies of basic cognitive abilities – of basic mentality – one finds in Levinas and Merleau-Ponty agrees with and lends further support to the radical EEC project. On these views, the fundamental sort of activity that minded beings engage in are not aptly described in representational terms and appears to be constitutively embodied and world-involving. The argument was not simply that, were one to describe one's perceptual grip on a surround, no representations would manifest themselves therein. That does seem to be a correct and non-trivial point, but it is clearly insufficient, when detractors can always take refuge in a flight to the subpersonal, where it may be claimed that the real – representational – action of cognition transpires. The phenomenology Levinas and Merleau-Ponty offer in favor of the radical EEC position is more than a witch hunt for representations in what some might take to be merely the phenomenal dross of cognitive processes.

The radical phenomenology we get from these thinkers also shows that representational explanations are both superfluous and ill-suited to what one can observe of cognitive activity at the person-level. One can ascertain certain gross structural and functional features of cognitive processes at the person-level. The ones that stand out for Levinas and Merleau-Ponty importantly pivot on the dynamic relation of conscious awareness to body and world in sensory experience. Levinas describes the attitude of enjoying and assuming that guides perception, which he views as a leap into the rich sensory array whose ignorance is precisely its virtue, just as long as that sensory activity is supported by a world that is more or less agreeable to it. Successfully, even intelligently, encountering a surround with one's senses need not disengage or defer to any higher-order cognitive process, but takes sensation for all it's worth. Crucially, the agreeableness of one's surrounding in sensibility, for Levinas, is a fit between body and world, since what stands out to sensibility, what it "devours" are those features of its surrounding that are of vital significance, that answer to its embodied condition. Merleau-Ponty takes the point further by emphasizing that the conscious guidance of perceptual activity works not only by relying on the world to support its activity, but to rely on itself – in all the complexity of its own body – to do the same. Merleau-Ponty views the body itself to be a complex synergetic system coupled to the world. The organization and coherence of cognitive processes, e.g., in perception, is a function of the dynamic interplay of these elements.

Let me summarize what I take to be the positive results of this radically embodied and enactive phenomenology. First, representations are supposed to help us identify items or features in our environment. Yet, sensory tendencies allow us to discriminate without modes of presentation or descriptions that aspects are supposed to fit, i.e., without semantic content. Second, representation is supposed to guide us prospectively in the precarious activity of navigating our surroundings. Sensory dynamics, though, need not maintain unilateral control or anticipate by way of expectation. Their success, in fact, is in large measure due to their being interwoven with embodied know-how and geared onto a cognitive niche to which they are adapted. Third, on representational views, goings on and factors primarily internal to the perceiver guarantee that perception takes shape as it should. Yet, on the radical picture sketched above, interactive dynamics take the lead instead of any single member of the mind-body-world system. Indeed, the (non-semantic) sense that arises in sensory experience belongs to the situation encompassing all these elements, not to the perceiver to which representations are supposed to belong. Overall, a uniquely perceptual, non-representational attitude is manifest in all of this at the level of conscious experience, an attitude that interweaves with body and world. Note, lastly, that body and environment, in cooperation with this non-cognitive attitude, manage in various ways to pick up a significant amount of the explanatory slack that representations are supposed to carry.

References

Anderson, M. 2014. *After phrenology: Neural reuse and the interactive brain*. Cambridge, MA: MIT Press.

Barandiaran, X., and A. Moreno. 2006. On what makes certain dynamical systems cognitive. *Adaptive Behavior* 14(2): 171–185.
Barandiaran, X., and A. Moreno. 2008. Adaptivity: From metabolism to behavior. *Adaptive Behavior* 16(5): 325–344.
Barrett, L. 2011. *Beyond the brain: How body and environment shape animal and human minds.* Princeton: Princeton University Press.
Beer, R. 1996. Toward the evolution of dynamical neural networks for minimally cognitive behavior. In *From animals to animats 4: Proceedings of the fourth international conference on simulation of adaptive behavior*, ed. P. Maes, M. Mataric, J.A. Meyer, J. Pollack, and S. Wilson. Cambridge, MA: MIT Press.
Beer, R. 2003. The dynamics of active categorical perception in an evolved model agent. *Adaptive Behavior* 11(4): 209–243.
Bower, M. 2014. Affectively driven perception: Towards a non-representational phenomenology. *Husserl Studies* 30: 225–245.
Brooks, R. 1991. Intelligence without representation. *Artificial Intelligence* 47: 139–159.
Buhrmann, T., and E. Di Paolo. 2014. Non-representational sensorimotor knowledge. *From Animals to Animats 13: Proceedings of the 13th international conference on simulation of adaptive behavior, SAB 2014*, LNAI 8575, 21–31. New York: Springer.
Burge, T. 1979. Individualism and the mental. In *Midwest studies in philosophy, volume 4, metaphysics*, ed. P. French, T. Uehling Jr., and H. Wettstein, 73–121. Minneapolis: University of Minnesota Press.
Carman, T. 2003. *Heidegger's analytic*. Cambridge: Cambridge University Press.
Chemero, A. 2001. What we perceive when we perceive affordances. *Ecological Psychology* 13: 111–116.
Chemero, A. 2003. An outline of a theory of affordances. *Ecological Psychology* 15: 181–195.
Chemero, A. 2009. *Radical embodied cognitive science*. Cambridge, MA: MIT Press.
Clark, A. 2008. *Supersizing the mind: Embodiment, action, and cognitive extension*. Oxford: Oxford University Press.
Clark, A., and J. Toribio. 1994. Doing without representing? *Synthese* 101: 401–431.
Crowell, S. 2013. *Normativity and phenomenology in Husserl and Heidegger*. Cambridge: Cambridge University Press.
Di Paolo, E. 2005. Autopoiesis, adaptivity, teleology, agency. *Phenomenology and the Cognitive Sciences* 4: 429–542.
Di Paolo, E. 2015. Interactive time-travel: On the intersubjective retro-modulation of intentions. *Journal of Consciousness Studies* 22(1–2): 49–74.
Dotov, D. 2014. Putting reins on the brain: Hoe the body and environment use it. *Frontiers in Human Neuroscience* 8(795): 1–12.
Dotov, D., L. Nie, and A. Chemero. 2010. A demonstration of the transition from readiness-to-hand to unreadiness-to-hand. *PLoSOne* 5: e9433. doi:10.1371/journal.pone.0009433.
Drummond, J. 1990. *Husserlian intentionality and Non-foundational realism: Noema and object.* Dordrecht: Kluwer.
Drummond, J., and L. Embree (eds.). 1992. *The phenomenology of the Noema*. Dordrecht: Springer.
Gallagher, S. 2014. In your face: Transcendence in embodied interaction. *Frontiers in Human Neuroscience* 8(495): 1–6.
Gallagher, S. 2008. Direct perception in the intersubjective context. *Consciousness and Cognition* 17: 535–543.
Gallagher, S., and D. Hutto 2008. Primary interaction and narrative practice. *The shared mind: Perspectives on intersubjectivity*, 17–38, ed. J. Zlatev, T. Racine, C. Sinha, and E Itkonen. Amsterdam: John Benjamins.
Gibson, J. 1979. *The ecological approach to visual perception*. Hillsdale: Lawrence Erlbaum Associates Publishers.

Golob, S. 2013. Heidegger on assertion, method, and metaphysics. *European Journal of Philosophy*, Published Online doi:10.1111/ejop.12018.

Hand, S. 2009. *Emmanuel Levinas*. London: Routledge.

Heidegger, M. 1962. *Being and Time*. Trans. J. Macquarrie and E. Robinson. New York: Harper Collins.

Heidegger, M. 1995. In *Fundamental problems of metaphysics: World, solitude, finitude*, ed. W. McNeill and N. Walker. Bloomington: Northwestern University Press.

Henry, M. 2008. *Material phenomenology*. Trans. S. Davidson. New York: Fordham University Press.

Husserl, E. (1982). *Ideas pertaining to a pure phenomenology and to a phenomenological philosophy: First book.* Trans. F. Kersten. Dordrecht: Martinus Nijhoff Publishers.

Husserl, E. 1983. *Ideas pertaining to a pure phenomenology and to a phenomenological philosophy, First Book.* Trans. F. Kersten. The Hague: Martinus Nijhoff Publishers.

Husserl, E. 1973. In *Experience and judgment: Investigations in a genealogy of logic*, ed. J.S. Churchill, K. Ameriks, and L. Eley. Evanston: Northwestern University Press.

Husserl, E. 1999. *Cartesian meditations: An introduction to phenomenology*. Trans. D. Cairns. Boston: Kluwer Academic Publishers.

Hutto, D. 2008. *Folk psychological narratives*. Cambridge, MA: MIT Press.

Hutto, D. 2011. Philosophy of mind's new lease on life: Autopoietic enactivism meets teleosemiotics. *Journal of Consciousness Studies* 18(5–6): 44–64.

Hutto, D., and E. Myin. 2013. *Radicalizing enactivism: Basic minds without content.* Cambridge, MA: MIT Press.

Levinas, E. 1969. *Totality and infinity: An essay on exteriority*. Trans. A. Lingis. Dordrecht: Kluwer Academic Publishers.

Levinas, E. 1987. *Collected philosophical papers*. Trans. A. Linguis. Dordrecht: Martinus Nijhoff Publishers.

Levinas, E. 1991. *Otherwise than being, or beyond essence*. Trans. A. Lingis. Dordrecht: Kluwer Academic Publishers.

Levinas, E. 1998. *Entre nous: On thinking-of-the-other*. Trans. M.B. Smith and B. Harshaw. New York: Columbia University Press.

MacAvoy, L. 2005. The other side of intentionality. *Addressing Levinas*, 109–118, ed. E.S. Nelson, A. Kapust, and K. Still. Evanston: Northwestern University Press.

Marratto, S. 2012. *The intercorporeal self: Merleau-Ponty on subjectivity*. New York: SUNY Press.

Merleau-Ponty, M. 1968. *The visible and the invisible*. Trans. A. Lingis. Evanston: Northwestern University Press.

Merleau-Ponty, M. 2003. *Phenomenology of perception*. Trans. C. Smith. New York: Routledge Classics.

Merleau-Ponty, M. 2010. *Institution and passivity: Course notes from the Collège de France*. Trans. L. Lawlor and H. Massey. Bloomington: Northwestern University Press.

Moran, D. 2014. What does Heidegger mean by the transcendence of Dasein? *International Journal of Philosophical Studies* 22(4): 491–514.

Morgan, M. 2011. *The Cambridge introduction to Emmanuel Levinas*. New York: Cambridge University Press.

Morris, D. 2004. *The sense of space*. New York: SUNY Press.

Orlandi, N. 2014. *The innocent eye: Why vision is not a cognitive process*. Oxford: Oxford University Press.

Putnam, H. 1975. The meaning of 'meaning'. *Minnesota Studies in the Philosophy of Science* 7: 131–193.

Reeke, G., and G. Edelman. 1984. Selective networks and recognition automata. *Annals of the New York Academy of Sciences* 426(1): 181–201.

Rowlands, M. 2010. *The new science of the mind: From extended mind to embodied phenomenology*. Cambridge, MA: MIT Press.

Smith, A.D. 2008. Husserl and externalism. *Synthese* 160: 313–333.
Smith, D.W., and R. McIntyre. 1984. *Husserl and intentionality: A study of mind, meaning, and language*. Dordrecht: Springer.
Thelen, E., and L. Smith. 1994. *A dynamic systems approach to the development of cognition and action*. Cambridge, MA: MIT Press.
Thompson, E. 2007. *Mind in life: Biology, phenomenology, and the sciences of the mind*. Cambridge, MA: Harvard University Press.
Van Duijn, M., F. Keijzer, and D. Franken. 2006. Principles of minimal cognition: Casting cognition as sensorimotor coordination. *Adaptive Behavior* 14: 157–170.
Van Gelder, T. 1995. What cognition might be, if not computation. *The Journal of Philosophy* 91(7): 345–381.
Vasseleu, C. 1998. *Textures of light: vision and touch in Irigaray, Levinas, and Merleau-Ponty*. London: Routledge.
Wilson, R. 2004. *Boundaries of the mind*. Cambridge: Cambridge University Press.
Wilson, A., and S. Golonka. 2013. Embodied cognition is not what you think it is. *Frontiers in Psychology* 4(58): 1–13.
Wrathall, M. 2011. *Heidegger and unconcealment: Truth, language, and history*. New York: Cambridge University Press.
Yoshimi, J. 2009. Husserl's theory of belief and the Heideggerean critique. *Husserl Studies* 25: 121–140.

Chapter 10
Mind and Value

Nathaniel F. Barrett

10.1 Introduction

The values of human experience, in the widest possible sense, encompass much more than those commonly singled out as moral and aesthetic values. They also include all varieties of enjoyment and interest, even objects of idle curiosity. At the lowest scale, anything salient enough to be registered in experience can be said to have some value. We cannot, therefore, mark off a set of value-free things in experience; we can only discern differences in how things are valued. Indeed, it is impossible to imagine experience without value. Whatever we might think about the world described by science, we cannot actually experience anything as just a collection of facts—utterly without perspective, affective tone, gradations of prominence, foreground or background. To capture this pervasiveness of value, rather than catalogue all the diverse values of experience, it is better simply to say that experience is valuation.

If experience is valuation, then it seems appropriate to ask: what is it valuation of? Do the objects of experience have values of their own, or does value only arise in experience? These question will occupy us in due time. For now, let us remain with the claim that experience is valuation and consider more closely what it entails.

Experience is valuation, first and foremost, because it is selective. What we actually experience is always a small fraction of what is available to be experienced at any given moment. Moreover, the selectivity of experience suggests valuation rather than delimitation because it is never all-or-nothing. The contents of experience are not sharply circumscribed, but fade into a vast penumbral background.

N.F. Barrett (✉)
Mind-Brain Group, Institute for Culture and Society (ICS),
University of Navarra, 31009 Pamplona, Spain
e-mail: nbarrett@unav.es

M. García-Valdecasas et al. (eds.), *Biology and Subjectivity*,
Historical-Analytical Studies on Nature, Mind and Action 2,
DOI 10.1007/978-3-319-30502-8_10

In fact the graded transition from distinctness to indistinctness is perhaps the most telling feature of experience as valuation, although it is often overlooked. At one extreme of this gradation is the foreground of experience, articulated by relatively clear and distinct contrasts. Contrasts include any and all ways of valuing certain differences as more important than others: contrasts are differences that matter. As such, we should not say that contrasts are wholly constructed, but it does seem that they are subjective at least insofar as their selection and enhancement of difference is made by and for a perspective.

In any case, the connection in experience between contrast and valuation is crucial. We are accustomed to thinking of values as somehow more subjective than the contents of experience—for instance, as affective feelings that are added to non-affective objects. But if the most basic contents of experience are constituted by subjectively formed contrasts, then valuation is fundamental: any recognizable object of experience already is a kind of valuation.

Unfortunately, when we think about the role of value in experience, we tend to forget its pervasiveness. What comes most readily to mind is the kind of value that stands out as such—the kind of value that seems to belong to a special sphere of experience and to attach only to objects of special interest. But while these especially prominent values are important clues to the nature of value in general, their specialness reinforces our tendency to overlook the fact that all experience, cognitive and non-cognitive, is valuation. Feelings of eminent worthiness do not emerge sui generis from a valueless background; they are intensifications of the very same valuations that constitute experience as such. Yet precisely because value is so basic to experience, it is easily taken for granted.

Moreover, even when the pervasiveness of value is admitted, there is a tendency to set it apart from the functions and causal mechanisms of cognition and lump it together with other problematically subjective phenomena. However, while value does have an ineradicable subjective and qualitative aspect, it is not reducible to *qualia*, at least not insofar as the latter are defined as *purely* subjective, *merely* qualitative phenomena, and therefore functionally and causally inert. Value is much harder to divorce from the various functional and causal relations of mind without introducing an artificial division within experience. This is not just because value in experience is inseparable from experiential content, as just noted, but also because the nature of experience as valuation or *valuing* carries with it a dynamic, relational dimension that begins with something valued, and therefore cannot be reduced to an isolated patch of qualitative phenomena. All feelings are valuations, and although they may not have a clearly defined object, they are always feelings *of* something, and cannot be severed from this objective pole.

Perhaps this "double-barreled" character of value—traversing subjectivity and objectivity—helps to explain the perennial philosophical controversies that have surrounded the nature of value. Questions about value extend back to the very beginning of western philosophy, but in the modern period they have been further complicated by their entanglement with the problem of consciousness. Indeed, one of the most unfortunate consequences of the modern extrusion of consciousness from nature has been the diminishment of value to a merely subjective property that minds project onto the world. Thus consciousness and value have suffered the same fate with regards to their place, or lack thereof, in the modern scientific worldview, as both have been banished to an inferior, quasi-non-natural realm. Recently, contemporary philosophers have attempted to put value and normativity back into nature (McDowell 1998), though few have focused on our experience of value as an approach to the problem of consciousness. However, if the preceding discussion is on track, understanding experience as valuation may provide us with a fresh angle on the gap between mind and nature.

For instance, we might consider how valuation in experience relates to physical causation in the wider world. Two major alternatives immediately present themselves: either causation does not involve valuation, in which case the problem is to understand the relation of valuation (in terms of emergence or reduction) to non-valuational processes; or causation does involve valuation, in which case the problem is to understand physical causation in terms of valuation, and to understand mental valuation as a distinct variety of causal process. The first option is basically just a different formulation of the mind-body problem; the second seems to do away with the gap between mind and body, but in its place it opens a whole new can of worms.

Now, a widespread presumption in modern philosophy and science is that physical causation, whatever else it may be, is *not* valuation. Indeed, few feel the need to make this explicit. The reasons for the apparent implausibility of the second option are not hard to fathom. As the preceding discussion has shown, if valuation occurs outside the mind, the subjective characteristics we associate with mind obtain elsewhere in nature. This smacks of panpsychism, the view that mind is ubiquitous in nature. While panpsychism has recently enjoyed a minor comeback within philosophy of mind (e.g., Strawson 2006), it has largely failed to win support as a solution to the mind-body problem. Perhaps the reason is that analytic versions of panpsychism tend to define the problematic subjective aspects of mentality in terms of some *property*—let us call it property X—and then extend this property to all natural entities. While this redistribution of properties deftly removes the gap between mind and nature, it does nothing to explain why property X exists in the first place (see Barrett 2009).[1]

As claimed above, however, valuation does not reduce to any particular qualitative property. Therefore the suggestion that valuation occurs outside the mind, and is perhaps intrinsic to causation, is not just an attempt to dismiss the Hard Problem of consciousness by ontological fiat. Nevertheless, for the second option to prove its worth, a concept of valuation that is derived from experience must be shown to improve our understanding of physical causation—that is, value must be shown to be important to our understanding of nature as well as mind. Needless to say, such an approach to the mind-body problem is much more ambitious than those that leave standard notions of physical causation in place (which is most of them). But that does not mean that it is not worth considering. Do we really think we can

[1]The failure of analytic panpsychism shows that the modern mind-body problem is not just an "explanatory gap" between two otherwise adequately known realms of experience.

get around the mind-body problem without addressing longstanding problems with causality?

The purpose of this chapter, then, is to make an initial case for the plausibility of the second option, drawing principally on the "speculative naturalistic" strand of modern philosophy that includes pragmatism and process philosophy, and for which (1) mind is valuation, (2) mind is continuous with the rest of nature so that (3) some kind of valuation also belongs to the rest of nature, and (4) the object of experience as valuation is also some kind of value. Given the speculative nature of this kind of approach to the mind-body problem, as well as its unfamiliarity for many readers, emphasis should be placed on the preliminary nature of the following discussion. What is presented here is an *approach*, not a *solution*: it describes the distinctive naturalistic and axiological orientation of American philosophy, especially as exemplified by John Dewey and Alfred North Whitehead, and more recently by the work of Robert Cummings Neville. In addition, the final section comments on the apparent compatibility between Neville's axiological philosophy of mind and nature and the dynamical view of cognition in neuroscience. Of course, an axiological interpretation of physical causation cannot be simply "read off" of natural phenomena, and so its justification requires much more than apparent compatibility. What explanatory advantages are gained, or at least promised, by this approach? One possibility, which can only be outlined here, is that an axiological theory of causation allows us to understand mind as a process of *engagement* with values, thus saving our notion of experience as a process of discovery.

But first, it is worthwhile to prepare the following discussion by examining the standard approach to value within cognitive science. Hopefully, by pointing out some the flaws of this approach it will be easier to motivate the turn to such a radical alternative.

10.2 Value in Mind: The Darwinian-Functionalist Approach

Today most scientific treatments of value trace its roots to functions that have arisen from processes of natural selection. Accordingly, the most basic values of human life are related to survival and reproduction. Even if neuroscientists readily admit that human experience is underdetermined by these basic values, the prevailing assumption is that our feeling of value—our experience of something as good, beautiful, or even just worthy of interest—is constructed by systems whose primary or original function is the pursuit of values determined by selection pressures in our evolutionary past (e.g. see Damasio 2005).

At first glance, it is difficult to say whether this Darwinian-functionalist approach can do justice to the phenomenology of value experience. The range of experiences that count as an experience of value is staggeringly large and diverse; indeed, according to the view just presented, it includes *all* of experience. Certainly we have no well-formed standards by which to judge phenomenological adequacy. However, problems become easier to detect once this approach is further specified by a cognitive theory.

A key implication of the Darwinian-functionalist approach is that value is (and therefore can be) encoded by specific structures of the human organism. First and foremost, values are encoded by the genome. Secondly, these "genetically programmed" values are elaborated in relation to environmental input by the more flexible "value systems" of the human nervous system (e.g. systems involved in emotion). Within this scheme, the role of value within cognition is typically understood as the appraisal of cognitive content, such that value is something *ascribed* to the inputs of cognition, and the function of value ascription is typically assigned to specific structures or "tiers" within the cognitive architecture (e.g., Grabenhorst and Rolls 2011, p. 57). In summary, according to this standard approach, value plays a subsidiary role in cognition, is encoded and ascribed by mechanisms that are distinct from those involved in information gathering and processing, and can be reduced to the functional roles that these mechanisms play within the larger cognitive process.[2]

Two problems arise from this definition of value within cognition. One is the problem of how to relate mechanisms of value ascription to an adequate phenomenology of value. Making the transition from value-encoding structures to experience seems to require that value shows up as a special class of qualities that are somehow intrinsically value-bearing and that attach selectively to objects of experience so as to make these objects appear as values or disvalues. Yet when values are directly experienced, they seem to be enmeshed with the whole qualitative and meaningful content of experience. Moreover, as the qualitative details of an experience are altered and enriched through perceptual interaction, so, it seems, are the directly enjoyed values. Indeed, rather than talk about values *in* experience as if they were discrete kinds of content, perhaps it is better to talk of the "value character" of experience. This value character does not seem to be a special qualitative overlay, something super-added to experience; rather it seems to be intrinsic to experience as such.[3]

The second problem is related to the first, although it is more cognitive than phenomenological. It has to do with the *grounding of value*, the problem of making sure that values are rightly ascribed to input received from the environment. When value is conceived as something added to input, the cognitive system as a whole can only recognize values as they are encoded by mechanisms within its architecture,

[2] This summary of the standard functionalist approach to value and some of the following criticisms draw from arguments for an alternative "enactive" approach presented in Rohde 2010, and Di Paolo et al. 2010. Accordingly, as an introduction to standard cognitive approaches to value, it may be biased. But it seems to accord with the scientific literature. See, for instance, Damasio 2005, and Grabenhorst and Rolls 2011.

[3] To this objection, it is possible to reply that since mechanisms of value ascription are unconscious, all that shows up in experience is always already clothed in value, such that value can be inseparable from phenomenological content and yet still be a product of a distinct stage of information processing. Yet this strategy saves the intricately variegated appearance of value character by placing an enormous computational burden on the evaluative substructures of cognition. A better strategy is to build value into our concept of mind, that is, to make valuation intrinsic to the cognitive process (and to experience) as a whole rather than assigning it a subsidiary functional role.

and these mechanisms of value-ascription, whose contribution is some kind of "value signal," can be easily rewired or otherwise corrupted (Rohde 2010). For example, with a simple "flip of the switch," what is valued as "food" might just as easily be valued as "poison" and vice versa (Di Paolo et al. 2010, p. 47). Such liabilities are not necessarily fatal, of course; they might be handled by additional systems of error detection.

But again, returning to phenomenology, the arbitrary manner in which a "value signal" functions seems to have no relation to the way value is experienced as grounded in engagement with the world. This is a delicate point, as a certain degree of "internality" and "perspectivity" in the experience of value cannot be denied. That is, when directly enjoyed, value does seem to possess a certain degree of autonomy. Therefore a distinction needs to be made between the intrinsic nature of value as something immediately felt, which cannot be mistaken, and processes of value interpretation and attribution, which can be mistaken. Yet we can uphold this distinction without isolating the direct experience of value as a simple, meaning-extrinsic quality that is "triggered" by input. Indeed, to account for the multifarious and highly nuanced value character of experience, it seems that our best option is to ground this experience in engagement—that is, the coupling of our bodies with a complex and changing world. This is the approach to which we now turn our attention.

10.3 Mind and Nature: The Speculative Naturalistic Approach

The term "speculative naturalistic" is perhaps a bit misleading, as the American tradition of philosophy to which it refers is not naturalistic according to the physicalist understandings of naturalism that currently dominate philosophy. The choice of the softer, adjectival form is therefore intentional, as there is no need to enter into broader issues of philosophical naturalism, such as the view that nature is completely self-sufficient, or to entertain any such claims about the totality of nature. The pertinent sense of "naturalistic" is much more modest: it signifies a commitment to the continuity of experience and nature, as well as a closely related way of doing philosophy—not so well defined as to constitute a method—that cycles between experience and nature.

In this respect, the speculative naturalistic stance resembles that of philosophers and cognitive scientists who adhere to a "circle of experience" that moves between phenomenological and scientific descriptions (Varela et al. 1991). Yet American speculative naturalistic philosophy is distinctive in at least two other important respects: its emphasis on aesthetic experience, and its willingness to undertake speculative forms of inquiry about nature, such as philosophical cosmology and metaphysics. What is important to understand is how these go together so as to support the extension of valuation to all of nature.

An enormous amount of relevant historical background must be left out of this account. For instance, the combination of naturalistic and aesthetic interests could be traced back through Emerson and Thoreau to Jonathan Edwards, who many consider to be the progenitor of this strand of American philosophy. And while it is hard to imagine a discussion about American philosophy in which Charles Peirce does not play a prominent role, for present purposes we will have to make do with a truncated exposition that focuses on how Alfred North Whitehead and John Dewey were both inspired by William James to elaborate their naturalistic philosophies out of the full breadth, depth, and variety of experience. Dewey and Whitehead exemplify American naturalistic philosophy in rather different ways but, for the most part, their differences are relegated to the background of this treatment. What is important is how their shared commitment to the continuity of experience and nature led to radical views concerning the place of value and valuation in nature.

Remarkably, much of what defines the common experiential orientation of Whitehead and Dewey can be found in the famous essay of William James, "The Stream Thought," first published in 1890 as part of his landmark work, *The Principles of Psychology* (1983). James's essay serves as our starting point because of its bold assertion—still pertinent today—that transitory, vague, indistinct, and otherwise "fringe" aspects of experience are just as essential to mind as the more stable and distinct aspects. It thus presents a clear and decisive break with the atomistic view of experience which became widespread after Descartes, and which was influential in helping to establish a positivistic view of scientific knowledge, a mechanistic view of nature, and both empiricist and rationalist schools of psychology.

Perhaps the most definitive feature of the atomistic view is the notion that experience is composed of "simple sensations" (1983, p. 219). This notion assumes that the relatively clear and distinct features of experience are also its most fundamental elements—its building blocks—such that experience can be analyzed into these features without remainder. Against this notion, James pointed out that such features, especially when separated from the rest of experience, are in fact derivative abstractions. In a sense, one could say that James's view of experience turns positivism upside down, making the presumed bedrock of experience into a highly refined product, yet its consequences for scientific knowledge are not so damaging as this upending would lead one to expect. On the contrary, as argued by Whitehead (1967a [1925]), only by expanding our view of experience to include its more elusive, analytically intractable features can we properly account for the possibility of scientific knowledge, for those who follow Hume in neglecting these features must also follow Hume in his devastating critique of the empirical basis for inductive knowledge.

More positively, "The Stream of Thought" forwards a number of claims about experience with important implications for speculative naturalistic philosophy. For instance, James asserts that experience is always changing, such that there can be no exact recurrence of sensations or ideas (1983, pp. 224–25). That experience involves change is obvious, but James makes the much stronger claim that experience is *never* exactly the same, even when it is terribly monotonous. If this is true, the consequences for the atomistic view and its epistemological aspirations are devastating: there are no stable "simples" on which to found a system of knowledge. Another claim is that experience contains transitory as well as substantive elements—"flights and perchings," as James puts it—linking the present moment with past and future (pp. 234–36). And a third claim is that thought is always a matter of selective emphasis, an accentuation of some features at the expense of others (p. 273).

In sum, James claims that experience is richly imbued with what we might now call "non-cognitive" features—traces, anticipations, backgrounds, horizons, layers, overtones, suffusions, inflections, shadings, fringes—that, in their own way, are just as essential to cognition as the content that is more readily grasped, analyzed, and formulated. Such features are essential because, inter alia, they supply us with what modern epistemology has assumed we cannot have: direct feelings of causation, brought to us by thc way in which our conscious experience is suffused with "warmth and intimacy" of our bodies (p. 235), as well as direct feelings of relation (pp. 237–38), including the relation between subject and object.

The radical nature of James's break with the Cartesian legacy of modern philosophy—including the philosophies of both Hume and Kant—cannot be overemphasized. However, it should be reiterated that James's position was not that experiences of causation and relation are reliably clear and distinct in the manner of, say, the normal perceptual discrimination of everyday objects. James's position rescues the larger causal context in which clear perception and reliable cognition are possible, but at the cost of making this larger context less than perfectly knowable. In other words, for James—as well as for Dewey and Whitehead after him—we come to know things only because experience always connects us with more than we know.

As pointed out in Thomas Alexander's fine study (1987, pp. 72–73), this Jamesian view is at the core of Dewey's philosophy of experience as it is elaborated in his most important works (1958 [1925], 1980 [1934], 1991 [1938]). Dewey's debt to James is especially clear in a relatively early essay, "The Postulate of Immediate Empiricism" (1997 [1905]), which states that "things…are just what they are experienced as" (p. 226). At first glance this seems like an impossibly naïve form of realism, and indeed it would take at least another two decades for Dewey to clarify his position. But Dewey's main point is that just because experience always presents us an imperfectly knowable world, we should not assume that this imperfect knowability is mere illusion due to our imperfect faculties, and that the real world is somehow constituted so as to afford complete knowledge to an all-seeing mind.

> By our postulate, things are what they are experienced to be; and, unless knowing is the sole and only genuine mode of experiencing, it is fallacious to say that Reality is just and exclusively what it is or would be to an all-competent all-knower; or even that it *is*, relatively and piecemeal, what it is to a finite and partial knower. Or, put more positively, knowing is one mode of experiencing, and the primary philosophic demand…is to find out *what* sort of an experience knowing is (pp. 228–29).

Note that in this statement is Dewey is not just talking about experience as more than knowing; he is also suggesting that nature is more than what is perfectly knowable. This is not obscurantism; Dewey does not limit knowledge by drawing a hard and fast line between knowable and unknowable parts of experience or nature.

Like Peirce before him, Dewey places no immovable barriers in the path of inquiry. On the contrary, one of Dewey's basic doctrines is that "*experience grows*, and in growing takes on *meaning*" (Alexander 1987, p. 80). A hallmark of pragmatism is the insight that in order to understand how we can and do make progress in our knowledge of the world we must abandon certain dearly held premises that would, if true, guarantee us the possibility of having *absolute* knowledge—whether absolute in completeness, comprehension, or certainty. One of the ironies of modern thought is that the false promise of this kind of knowledge has typically been purchased by a viciously impoverished account of experience. Dewey's principal debt to James, then, is the latter's replenishment of experience, which recovers many of the features that were jettisoned for the sake of perfect knowledge and yet are necessary for its growth and enrichment. But Dewey was also concerned to use this fuller notion of experience to understand nature, including, especially, nature as the broader context in which all varieties of experience and their capacities for growth are possible.

The overall trajectory of American naturalistic tradition of philosophy is therefore a reversal of what can be crudely summarized as a two-step process that begins with Descartes' view of experience and culminates in the triumph of Newtonian natural philosophy: the first step was the assertion that to gain certain knowledge experience must be analyzable into discrete, perfectly knowable elements, and the second step was the extension of similar standards of perfect knowability to the whole of nature. In James's "Stream of Thought" we find a decisive rejection of the premise that experience is so analyzable, and in the philosophies of Dewey and Whitehead we find the much more laborious (indeed, largely unfinished) process of working out what nature could be, given all that experience is.

In one of his last essays, Dewey singled out this latter project, that of elaborating a philosophy of nature that attempts to do justice to the full "background" of experience, as constituting a deep affinity between his philosophy and Whitehead's (1988 [1941]). He describes this affinity as a kind of philosophical method rather than a set of ideas or theoretical commitments, but it is something less formalized than what "method" implies. In Dewey's words, he and Whitehead share

> "…the ideas that experience is a manifestation of the energies of the organism; that these energies are in such intimate continuity with the rest of nature that the traits of experience provide clues for forming 'generalized descriptions' of nature…and that what is discovered about the rest of nature (constituting the conclusions of the natural sciences) provides the organs for analyzing and understanding what is otherwise obscure and ambiguous in experience directly had…." (p. 125).

In terms of philosophical style, however, the difference between their respective philosophies of nature is striking: whereas Dewey preferred to extend everyday language as far as it could go, Whitehead developed his own abstruse terminology. Also, Whitehead's philosophy, especially as represented by his magnum opus, *Process and Reality* (1978 [1929]) is much more ambitious in both speculative and systematic terms: it offers a categoreal scheme for the interpretation of all experience, including all scientific knowledge. However, while Whitehead's categoreal scheme aims at necessity, it is not necessary in the sense of proceeding by necessary deductions from self-evident premises. Rather, it is a hypothesis derived from experience and whose validity lies in its service to the growth and elucidation of experience. In the development of philosophical categories, Whitehead argued that we must take account of "every variety" of experience, including its darkest recesses:

> Nothing can be omitted, experience drunk and experience sober, experience sleeping and experience waking, experience drowsy and experience wide-awake, experience self-conscious and experience self-forgetful, experience intellectual and experience physical, experience religious and experience skeptical, experience anxious and experience care-free, experience anticipatory and experience retrospective, experience happy and experience grieving, experience dominated by emotion and experience under self-restraint, experience in the light and experience in the dark, experience normal and experience abnormal (1967b [1933], p. 226).

In this respect, Whitehead and Dewey were deeply aligned, as Dewey himself noted (1988).

Moreover, as suggested by the passage just quoted, the notion of experience that informs Whitehead's philosophy is equally Jamesian, if not more so. In fact it seems that many of the phenomenological claims laid out in James's "Stream of Thought" were deliberately adopted by Whitehead as elements of his philosophy of organism (1978). Among these is the rejection of doctrine that experience is reducible to "clear-cut data," as well as extensive arguments in favor of the Jamesian view that every occasion of experience is unique, temporally thick, and suffused with a sense of intimate bodily connection that carries with it a vague sense of wider physical impingements. Thus every experience constitutes a perspective that recedes from a focal area of relatively clear and distinct contrasts into a vague horizon, extending indefinitely into the rest of nature (e.g., 1967b, pp. 225–26). Indeed, one could argue that James's view of experience is brought to its fullest expression in Whitehead's philosophy, in terms of faithfulness to the depth and variety of experience, the consequences of this view of experience for modern theories of perception and knowledge, and its implications for philosophy of nature.

Consider, for instance, Whitehead's defense of "perception in the mode of causal efficacy," his term for Jamesian feelings of relation and causation. Whitehead points out that if Hume is right that the perception of causation arises only from the habitual association of sense data, then feelings of causation should be absent when the normal use of sense organs is temporarily impaired.

> Unfortunately the contrary is the case. An inhibition of familiar sensa is very apt to leave us a prey to vague terrors respecting a circumambient world of causal operations. In the dark there are vague presences, doubtfully feared; in the silence, the irresistible causal efficacy of nature presses itself upon us; in the vagueness of the low hum of insects in an August woodland, the inflow into ourselves of feelings from enveloping nature overwhelms us; in the dim consciousness of half-sleep, the presentations of sense fade away, and we are left with the vague feeling of influences from vague things around us. It is quite untrue that the feelings of various types of influences are dependent upon the familiarity of well-marked sense in immediate presentment. Every way of omitting the sensa still leaves us a prey to vague feelings of influence. Such feelings, divorced from immediate sensa, are pleasant or unpleasant, according to mood; but they are always vague as to spatial and temporal definition, though their explicit dominance in experience may be heightened in the absence of sensa (1978, p. 176).

Again, along with the emphatic rejection of Hume's Cartesian suppositions, it is important to note that, as indicated by Whitehead's repeated use of the word "vague," feelings of causal efficacy, even at their most powerful, are never present in the clear and distinct manner of normal sensory perception, especially as the latter is exemplified by vision. Hume is right, after all, that we do not *see* causal relations or powers. Thus when Whitehead talks of perception in the mode of causal efficacy, he is not pointing out some neglected venue of scientific observation. For the collection of reliable scientific evidence, Whitehead readily concedes to positivism: only the clear and distinct deliverances of our sense organs will suffice. For instance, we cannot develop a workable theory of gravity from our "vague feeling of influences from vague things around us." The point is that when experience is reduced to "clear-cut data," we lose something essential for the adequate interpretation of that data, namely feelings of continuity, relation, and causal connection to the wider world.

It may seem that the vagueness of these feelings threatens to prevent us from reaching our overall aim, which is to indicate the experiential grounds for treating physical causation as valuation. But the preceding discussion has indicated that the expectation of *clear and distinct* experiential grounds for this or any other view of causation is unfounded. For such fundamental questions, the best that experience can offer is the suggestion of hypotheses, which must be refined and elaborated and then tested against experience in various modes. Yet we have not come up empty handed, for the features of experience to which the preceding discussion has repeatedly called attention offer some support for the theory that experience is valuation and, what's more, they illustrate at least some of what experience as valuation concretely entails.

Now, given that every occasion of experience is also an instance of causation, a fuller account of experience as valuation should supply us with a richer set of possibilities for theorizing about natural processes in general. This premise is entailed by the principle of continuity, which, according to Dewey, "authorizes us, as philosophers engaged in forming highly generalized descriptions of nature, to use the traits of immediate experience as clues for interpreting our observations of non-human and non-animate matter" (1988, p. 127). And so, though we cannot go much further into the details of either philosophy of nature, Whitehead's or Dewey's, let us draw out a few implications related to causation as valuation.

Implicit within the preceding discussion is the suggestion that vague feelings of relation are characteristic of experience as valuation. Considered a bit more carefully, this suggestion seems to hold up: if not for the transition between focus and fringe, feelings of relation would be much easier to point out, but they would not be valuations. That is, if the felt relations of experience were not partly vague, they would simply be relations, not valuations. Therefore, to eliminate vagueness from relationality in experience is also to eliminate valuation—and perhaps this goes for the rest of nature as well.

But what does it mean to talk about vagueness *in* nature, even in the most speculative terms? Simply put, vagueness is intrinsic to valuation insofar as valuation constitutes relations of selective emphasis and de-emphasis, or relations of differential importance. Valuation involves bringing something forward as important for an individual perspective, while simultaneously leaving something else behind as vague and trivial. Therefore, if a relation does not involve such differentiations of importance, it should not be considered as valuation. Perhaps the converse is also true: if the causal determination of an event involves a unique set of relations of differential importance, then it constitutes a kind of valuation. In fact the theory of causation forwarded by Whitehead entails something like this claim, as well as a genetic account of the "decision" that determines this unique set of relations.

Another clue from the preceding account of experience is that the continually changing nature of experience is also due to its character as valuation: each moment of experience is uniquely defined as an individual valuation of its causal context in the immediate past. Because of the nature of valuation, the differences between one moment and another may be so trivial as to be entirely negligible—that is, what is carried forward as important from moment to moment may be the same. Nevertheless, it should be remembered that any element of experience that we treat as literally unchanging is an abstraction from the totality of experience (which is not to say that it is not real). Likewise, perhaps the "bits of matter" that modern science has posited as the unchanging elements of our universe might best be treated as abstractions from a continual flux of events that, for the very same reason of relational constitution, never exactly recur. This is precisely the sort of argument that Whitehead made in his classic analysis of modern science (1967a [1925]).

Admittedly, Whitehead's cosmology is a tough swallow, especially for philosophers that have grown up in the relatively anti-speculative ethos of analytic philosophy. Yet many of Whitehead's seemingly outlandish ideas have a way of turning up in the writings of other philosophers and scientists. For instance, consider the preceding claim that no two causal events are exactly the same. The uniqueness of causation as valuation depends in part on assuming that each event draws from a unique context, which for Whitehead encompassed the entire actual world in the immediate past of the event in question. This would imply that the full determination of an event, qua valuation of the entire actual universe, is impossible for us to know or even come close to knowing—although perhaps we can reliably describe how certain events are determined because, for a great many events, most of this context can be disregarded as trivial. Now, as outlandish as this image of whole-world causality might seem, it bears a striking resemblance to the ideas recently floated by the physicist Lee Smolin in his recent manifesto, *Time Reborn* (2013). Therein, Smolin mounts the case that any physical description that limits itself to a particular region of the universe—what he calls "physics in a box"—is necessarily an approximation, and that in fact every particular event is "asymmetric" in the sense of constituting a unique set of relations. So, while the idea that each causal event constitutes an individual perspective on the entire universe is highly speculative, it is not without relevance to the contemporary world of physics.

Now, having fleshed out the nature of valuation in terms of differential relations of importance, this introduction to the American naturalistic approach requires the addition of one last but crucial point. In the opening paragraphs of this section, it was claimed that this strand of philosophy is characterized by the combination of a naturalistic orientation with an emphasis on aesthetic experience, and yet the latter has not been mentioned since. So what needs clarification is the role of aesthetic experience in this account of valuation in experience and nature.

For Whitehead as well as Dewey, aesthetic experience is not limited to a special form or sphere of experience: it is the epitome of experience qua valuation, and as such presents an essential clue to the nature of experience in general as well as the nature of valuation in general. Aesthetics in this broader sense refers to the enjoyment of intrinsic value or worthiness, as discovered in some object or event, or perhaps as found in the act of experiencing itself. It could be argued that any phenomenology that does not account for this dimension of direct enjoyment is inadequate, but certainly any account of experience qua valuation must give it special consideration.

Once attention is drawn to aesthetic experience, we can see that it supplies an important missing piece to the preceding account of valuation. So far we have focused on valuation as a kind of relation characterized by selective emphasis. Yet what determines this selection? As long as there are multiple possibilities, valuation is not just the differential importance of things for a particular perspective, but a particular *way* of taking up a perspective, and that begs the question: Why value something one way and not another?

If we accept that some event or occasion of experience constitutes a kind of valuation, we are led to demand a full account of that valuation that includes its reasons for valuing things one way and not another. Moreover, if every event qua valuation is a unique individual, the ultimate reason for a valuation cannot lie outside of itself. It has to be given by the intrinsic value that it attains by nature of its self-determined identity. Of course, practically speaking, not all valuations in human experience demand such an account. As we will see in the next section, most justifications of mental valuation can rest with an account of the context and purpose of valuation. But sometimes purposes themselves need accounting for, and aesthetics is the realm in which purposes, or ends, are considered in and of themselves. Peirce argued that logic is a species of ethics, and ethics is a species of aesthetics: that is, logic concerns a special kind of good, ethics concerns all that aims at some kind of good, and aesthetics concerns goods in themselves (Neville 1989, p. 49). All inquiry, therefore, leads ultimately to aesthetics, "the study of what is intrinsically worthwhile" (ibid.)—or else it leads to arbitrariness.

What, then, is the significance of aesthetic experience for our understanding of nature? This is perhaps the deepest question posed by the American naturalistic tradition. In Dewey's philosophy, the question is explored largely within the context of life: as stated above, Dewey saw experience as a manifestation of the "energies of the organism," and in his definitive statement of the aesthetic nature of experience, *Art as Experience* (1980 [1934]), he begins with a consideration of the roots of aesthetic enjoyment in the basic homeostatic rhythms of life. In Whitehead the question is pushed much further, to the point of suggesting that some rudimentary form of self-enjoyment is operative in all physical processes.

Whitehead's technical term for the physical germ of aesthetic experience is "intensity" (1978, pp. 83–85), which he considers basic enough to be included as part of his categoreal scheme (p. 27). An exposition of the central role of intensity in Whitehead's philosophy of nature is well beyond the scope of the present argument (but see Jones 1998). For now, let it suffice to note that the principal measure of intensity is contrast—not just sharpness of contrast, but also variety and richness of contrast. Recall that contrast was first introduced above, in the opening discussion of experience as valuation. Now contrast returns as a candidate for the definitive feature of intrinsic value.

Perhaps we are not accustomed to thinking of contrast as the mark of value, let alone intrinsic value. As an illustration of its relevance to value, think of how the aesthetic enjoyment of a lively conversation among good friends is characterized by a kind of easy flowing yet energetic intensity that is readily distinguished from the crude intensity of an angry confrontation. Contrast is a prominent ingredient in both kinds of intensity, but the richness of contrast that characterizes the former kind is what characterizes value at the highest scales. Contrast-rich intensity is achieved by gathering together different components—such as the different personalities and opinions among a group of friends—into a single pattern so that their differences are enhanced and vivified by joint participation in a complex, multidimensional harmony. In the case of an angry confrontation, most component differences are subsumed by a single dominant contrast that, while extremely intense, results in an overall loss of contrast, and thus an overall loss of intensity.

The singular advantage of contrast is that it can be understood both as a basic feature of any conceivable form, structure, or pattern *and*, because of its role in the enhancement and enjoyment of difference, as a basic dynamic and relational feature of experience. Contrast bridges the living waters of experience with the structures and formal properties of things. Thus, within the American naturalistic tradition as it has developed after Dewey and Whitehead, contrast is emerging as a crucial link between consciousness, the body, and the wider world, as the following section will attempt to make clear.

10.4 An Axiological Interpretation of Neurodynamics

The goal of this section is to show how the axiological philosophy of Robert C. Neville (1981, 1989, 1995), perhaps the greatest living exponent of the American speculative naturalistic tradition, can be used to interpret neuroscientific research pertaining to complex brain dynamics. Like the discussion of Dewey and Whitehead's philosophy in the previous section, this application of Neville's philosophy to neurodynamics is necessarily a cursory one, leaving behind a trail of loose ends. The aim is to show how the view of experience as valuation can help us to understand the brain as the primary organ of cognition.

In recent years, it has become fashionable to argue that to understand human cognition we have to get out of our heads and bodies (Noë 2009; Barrett 2011).

Meanwhile other theorists have argued that mind is not limited to animals with complex nervous systems, but should be extended to all forms of life (Thompson 2007). These are important arguments, and they show how well cutting-edge neuroscience fits with the century-old pragmatist approach to mind and knowing (Schulkin 2008). Yet no one would deny that the brain is a central piece of the puzzle that is the human mind.

Indeed, one could make the case that in recent years the brain has become even more puzzling than ever, and that this has happened precisely because of certain notable advances in neuroscience. The reason is that, while we have long known that the brain is staggeringly complex, our understanding of the nature of this complexity is changing.

At least until a few decades ago, our picture of the brain was dominated by its *structural* complexity. The preoccupation with structure was likely due in part to the kinds of investigative methods that were available, but it also stemmed from a particular view of how the brain performs its seemingly unique functional role as the seat of cognition: that is, the brain has generally been thought to be the cognitive organ par excellence in virtue of its highly intricate structure. Thus structural complexity has also undergirded the notion that the primary goal of cognitive science research is to describe the distinctive human "cognitive architecture." This is not to say that cognitive scientists have closely identified our cognitive architecture with anatomical structures; rather the concept of cognitive architecture applies first and foremost to a set of formal or logical structures. Yet the conviction that our distinctive cognitive capacities can be described as a system of formal or logical structures has been bolstered by the conviction that the massive and intricate anatomical complexity of the brain is sufficient to embody these logical structures.

The problem raised by neuroscience research of the past few decades is that it has added a whole new layer of complexity to the brain, namely *dynamical* complexity (Sporns 2011). In short, what is problematic about the dynamical complexity of the brain is that, unlike structural complexity, its patterns are constantly changing. Of course, the anatomical structures of the brain change too, but at a rate much slower than cognitive activity. What is meant by dynamical complexity is the fact that patterns of brain activity are highly transient and variable, emerging and dissolving at a rate that calls the idea of a single, species-general cognitive architecture—comprised of a massive hierarchy of functionally specialized structures—into question.

In fact, neuroscientists have known for decades, if not much longer, that the dynamic patterns of brain activity are highly irregular in the sense that the exact same pattern never appears twice. (William James argued this very point in his *Principles of Psychology*, for instance.) Yet the implications of this finding have been slow to dawn on the cognitive theoretical wing of neuroscience: how do we understand human knowledge in light of the fact that the principal organ of cognition never enters the same state twice? This question is easily dodged as long as there is room to disagree about what kind and scale of brain activity constitutes the neural basis of cognition, and there seems to be plenty of room. Certainly one can find an abundance of regular, patterned activity in the brain. At the level of the individual neuron, for instance, patterns of regular responses to specific stimuli seem to be well established (Hubel and Wiesel 1959; but see Chirimuuta and Gold 2009), and this fact fits well with the view that neurons, much like the logic gates of a computer, are the basic units of our cognitive architecture. Meanwhile, other kinds of regular activity can be measured at much larger scales: for example, the study of patterns of increased blood flow in different areas of the brain is one of the most prominent kinds of neuroscience research today.[4] Accordingly, depending on what one takes the most relevant brain activity to be, the standard architectural picture of cognition can seem fairly stable, and irregular brain activity can be dismissed as background noise.

However, an increasing number of neuroscientists are taking up a very different view of the most relevant activity for cognition, focusing on the large-scale, self-organized or "intrinsic" dynamics of populations of neurons numbering at minimum in the tens of thousands (Kelso 1995; Edelman and Tononi 2000; Cosmelli et al. 2007). Those who take this "neurodynamic" approach tend to embrace irregularity and regularity, chaos and order, as equally important and functional aspects of brain dynamics, calling into question the architectural view of the brain's role in cognition. Perhaps the neuroscientist who has gone farthest in this direction is Walter J. Freeman (1991, 1999).

Beginning in the 1970s, Freeman conducted EEG studies of olfactory bulb activity in rabbits as they were exposed to various "oderants" or olfactory stimuli. Again, it is crucial to specify the activity of interest: Freeman's olfactory research focused on patterns of amplitude modulation (AM patterns) found in the common waveform of the dendritic potential that spreads across the entire olfactory bulb. This is not the kind of neural activity that is typically presented as the basis of cognition in neuroscience textbooks. But for reasons that should soon become clear, it may be ideally suited to the ever changing, highly situational cognitive demands of real life. The full picture that emerges from Freeman's findings is rather complex (1999), but its basic import—and the challenge it presents to our understanding of the brain as the organ of cognition—can be summarized as follows.

First, Freeman found evidence that global AM patterns, rather than specific cells or networks, serve as the neural basis for the categorization of oderants. This means that while individual bulbar cells are linked with specific receptors and are therefore individually stimulated by oderants, olfactory *perception* involves, at minimum, *the entire olfactory bulb*, via an emergent waveform of the dendritic potential in which *every* neuron of the bulb participates—not just those stimulated by receptor activity (1999, pp. 72–74). Global AM patterns are constituted by local variances of amplitude in this common waveform, and these patterns can be detected by EEG readings and compared across different stimuli and repeated exposures to the same stimulus.

As interesting as this phenomenon of self-organized AM patterns may be, however, Freeman's most fascinating discovery concerns *how* these patterns function as perceptual categorizations. In short, Freeman's research shows that these patterns

[4] Importantly, most brain imaging studies average over many trials and multiple individuals, filtering out the variability of activation patterns.

are much less stable and regular than one would expect. Each rabbit has its own AM pattern signature that evolves out of that rabbit's individual history of stimulation (p. 76). Also, because of what Freeman calls "perceptual drift," the AM patterns of the same individual are not consistent enough to yield a stable set of averages, frustrating the expectation that oderants are categorized by building up and storing an average pattern for each oderant. Furthermore, when a rabbit is trained to identify a new oderant, the development of a new AM pattern for this oderant also changes all the other AM patterns as well (p. 81): that is, adding a new category changes the entire "category space." But even when no learning occurs, the repertoire of AM patterns in the olfactory bulb is constantly changing—because of perceptual drift, because of the current state of arousal or interest, because of the simultaneous presentation of other stimuli, or even just because of the order in which stimuli are presented. As a result, the AM pattern that results from the presentation of a particular oderant is determined as much by the current intrinsic dynamics of the olfactory bulb as it is by the stimulus, similar to the way the effect of a new arrival on a social gathering depends as much on the current dynamics of the gathering as it does on the newcomer. It seems, therefore, that there is no way to create a stable mapping between oderants and patterns of bulb activity.

For most cognitive theories, including most if not all varieties of computational theory that currently prevail within cognitive neuroscience and philosophy of mind, these results should be shocking. If something like these AM patterns is the relevant activity for such basic cognitive functions as perceptual categorization, what should we make of their lack of invariance with respect to stimuli (p. 92)? By Freeman's own account, it took him many years to come to grips with what his data was telling him about the dynamics of perceptual categorization, as he kept trying to interpret bursts of neural activity as if they were "inner representations" (Freeman and Skarda 1990). Indeed, the questions raised by Freeman's research go deeper than modern cognitive theory, as they touch on the basic issue of the correspondence between mind and world that has been debated by philosophers since Plato and Aristotle. The classical position is that knowledge is constituted by some kind of formal correspondence, so that the mind takes on the form of the object. What can we say, given the implications of Freeman's results, about the possibilities of any such correspondence?

It depends, of course, on what is meant by *form*. We know that the transmission of energy from objects of perception to our sense organs involves transformations of structure: for example, in the ear, vibrating air is transduced by receptors into bursts of electrochemical activity. Yet the common sense position is that, for knowledge to be possible, some kind of correspondence must obtain across these transformations, so that forms resident in the world are captured by corresponding forms of the perceiving organs. Thus the hope for commonsense realism would seem to rest on uncovering a fairly consistent mapping between structural features of the perceptual process and structural features of the world, and it is precisely this hope that is challenged by Freeman's account of neurodynamics.

Moreover, the challenge presented by neurodynamics to classical correspondence theory is not new. Freeman's work on the olfactory bulb is only an especially vivid empirical demonstration of dynamic features of perceptual categorization that have become increasingly apparent over the past half-century. In the case of visual perception, as argued by Francisco Varela, Evan Thompson, and Eleanor Rosch in their landmark work, *The Embodied Mind* (1991), the standard "opponent processing theory" of color vision (now decades old) presents a clear challenge to commonsense realism insofar as the latter depends on some kind of structural correspondence between mind and world (pp. 157–71). Just as patterns of olfactory bulb activity do not map onto different oderants, the perception of color does not map onto the wavelengths of ambient light or the reflectance properties of surfaces. As described by Varela et al., colors are contrasts that are "drawn forth" by the visual system (ibid.); colors are therefore ineluctably interpretive in the sense that they cannot be grounded independently in the world.

On the other hand, the contrasts of color vision are not just subjective projections, either (p. 170). The visual system *does* register differences of light wavelength through its receptors, just as the olfactory system registers differences of chemical structure through its receptors. Yet, similar to the way color contrasts are "drawn forth" from patterns of stimulation, Freeman's research suggests that categories of the olfaction are "drawn forth" by the entire olfactory system: in either case, perceptual categorization is a selective process of differentiation, drawing forth *distinctions or contrasts* out of differences of stimulation, rather than the detection of specific stimuli.

Thus one could say that the difference between individual receptor activity (i.e. sensory stimulation) and the activity of the perceptual system as a whole (i.e. perception) has to do with the way in which certain differences of the former are *valued* by the latter. Importantly, to *value* differences in a certain way is not the same as *constructing* them out of nothing, and too often this distinction is lost in discussions of how the brain "makes" the world. Insofar as they turn elaborate patterns of stimulation into contrasts which are meaningful to the organism, perceptual systems, as well as the entire brain and nervous system, should be considered as organs of valuation, not construction.

Granted, it is misleading to say that receptors value their stimuli; perhaps it is better to say that they are simply *triggered* by any appropriate stimulus that rises above a certain threshold. But the mechanistic notion of "triggering" should not be applied to the level of a perceptual system or the brain as a whole. As suggested by the phenomenon of color vision and by Freeman's studies of olfaction, perceptual categorizations consist of contrasts that depend on how differences of stimulation are *taken up* by the intrinsic activity of the perceptual system. Depending on various contextual factors such as arousal and past history, the differences of receptor activity that are "triggered" by specific oderants may or may not be registered as significant contrasts by distinctive patterns of bulb-wide activity. Furthermore, patterns of bulb-wide activity function as perceptual categorizations only insofar as they contrast with other patterns within the bulb's current category space or "contrast space" (Juarrero 1999, pp. 181–87).

Again, to be clear: in the case of neurodynamics, we are not simply talking about patterns of network activity that produce categorical distinctions by generalizing over patterns of stimulation. If that were the case, neurodynamic models like Freeman's would be no different from standard connectionist models. What makes neurodynamics different—and more challenging to commonsense realism—is not just the relative transience of its patterns, but how these patterns constitute meaning: namely, by forming a behaviorally significant contrast within the dynamics of the entire perceptual system. This dynamics is shaped by the animal's individual past history and current state of arousal *as well as* the incoming stimulus. And its meaning has to do with the potential consequences of the resulting contrast within the dynamics of the perceiving system and as well as the wider dynamics of the organism plus environment. Thus, for Freeman's model, *the meaning of a stimulus depends on how the animal acts so as to receive that stimulus*. An act of perception, therefore, is essentially a valuation of a stimulus.

For standard connectionist models, meaning is carried by patterns of "output" activity, each of which constitutes a categorization of some range of "inputs." Such patterns are more tightly wedded to structural features of the network: they depend on the arrangement and relative strength of connections between the nodes of the network, and these features change only gradually through learning. Consequently, what they mean is not a product of an active valuation by the perceptual system—at least not at the timescale of perception itself. Granted, one could make the case that the distinctive connectivity of a network constitutes a kind of valuation insofar as the learning process is actively guided by the animal's interests. But in response to a specific burst of stimulation, the network functions passively: it simply converts this "input" into an "output." Accordingly, the meaningfulness of network (as opposed to neurodynamic) activity seems to require a stricter kind of formal correspondence: patterns of network activity should somehow map onto patterns of stimulation, which in turn somehow map onto structural features of the world. Yet it seems that this kind of correspondence does not hold.

Therefore the difference between active valuation and passive transduction offers an important clue for understanding the functional roles played by different kinds of neural activity. As discussed earlier, it seems that many components of the nervous system function in a relatively passive manner: e.g., receptor cells seem to be "triggered" by specific kinds or levels of stimuli. While individual neurons are never purely passive conduits—for example, their readiness to fire can be modulated—we can roughly discriminate degrees of passivity along a continuum, according to range and timescale of the variability of response. The more transient and variable a kind of receptivity is, the more actively it can be shaped by the animal according to the needs of the situation, and the more fully it constitutes an active valuation of a stimulus rather than a passive transduction.

As exemplified by the AM patterns of the olfactory bulb, then, the key feature of perceptual neurodynamics is its "proactive" approach to stimulation by the active production of a transient global state of "preafferent" intrinsic activity that receives stimuli in a specific way (Freeman 1999). When combined with stimulation, this intrinsic activity settles into a pattern that constitutes a meaningful *contrast* with other states. Cognitively speaking, the advantage of this approach is that the meaningfulness of contrasts is highly individualized and situation-specific, as it is always a function of the unique history, state, and interests of the animal. In other words, one could say that by putting more individual and situational factors *into* the act of valuation, the animal is able to get more meaning *out of* the resulting contrast.

The challenge for theoretical understanding is to articulate how these contrasts are grounded in the world so as to constitute reliable knowledge. Contrasts are evidently highly perspectival, individualized phenomena, but that does not necessarily mean that they are "purely subjective." Is it possible, then, that some kind of correspondence might obtain at the level of contrast?

A rigorous argument for such a possibility is presented in the second volume of Robert C. Neville's trilogy on the axiology of thought, *The Recovery of the Measure* (1989). We have already set the stage for a brief consideration of Neville's thesis that truth is the "carryover of value" through the preceding analysis of the dynamics of perception as an active valuation of patterns of stimulation for which the contrasts embodied by patterns of neural activity, rather than the patterns per se, are the effective carriers of meaning. The apparent problem for realism is that such contrasts, qua products of fairly individualized, context-specific valuations, cannot "mirror" reality in any straightforward sense. Yet just because the mind cannot be said to mirror nature does not necessarily mean that concepts of reality and truth must be abandoned (cf. Rorty 1979). Neville's metaphysical brand of pragmatism develops a philosophy of nature, including a theory of causation as valuation, in order to defend a form of realism.

As indicated by its subtitle, "Interpretation and Nature," one of the central claims of Neville's book is that human interpretation is a refinement of causal processes that are ubiquitous in nature. Thus Neville stands squarely in the American speculative naturalistic tradition that stems from Peirce, James, Dewey, and Whitehead. While most modern philosophers have taken the thoroughly interpretive character of experience as a kind of epistemological confinement from which we must either gain non-interpretive access to reality or else give up on reality altogether, Neville rejects the idea that interpretation sets us apart from the rest of the natural world—but without reducing interpretation to something purely mechanistic. Only by developing a non-reductive, naturalized account of interpretation, Neville argues, can we understand how the world participates *causally* in our experience without reaching outside interpretation.

Thus Neville's approach can be seen as classical realism with a late-modern twist, preserving the possibility that experience is measured by reality while acknowledging the complications that have been brought to light by modern hermeneutics. On the one hand, Neville follows Aristotle in affirming the dyadic nature of reality—"reality is what it is and not some other thing" (p. 25)—as well as the dyadic nature of truth: "To say of what is that it is not, or of what is not that it is, is false, while to say of what is that it is, and of what is not that it is not, is true; so that he who says of anything that it is, or that it is not, will say either what is true or what is false" (p. 26, quoting Ross's translation of Aristotle's *Metaphysics* at 1011b 26). On the other hand, following the pragmatist Charles Peirce, Neville argues that the dyadic relation of truth is always embedded within a triadic (or even tetradic) framework of semiotic interpretation consisting of object, sign, and interpretant (more on this below). Thus one might say that, because of this embedding, while we can grasp the truth, we can never be completely certain that we have done so.

Neville's concession to interpretation may seem to leave us trapped within the "hermeneutical circle." But a key claim of his argument is that interpretation qua Peircean signification is a kind of causal process in which the thing interpreted participates as one of the causes. Moreover, he holds that the causal participation of the interpreted thing entails its actual presence in experience, albeit in some mediated form. The possibility of such presence is defended by Neville as a necessary condition for recovering the world as the measure of our interpretations, but it does not guarantee easy measuring—in some cases, the exact nature of the intended object's participation in experience may be very difficult to pinpoint. Nevertheless, compared to the European hermeneutical tradition, Neville presents a very different view of the challenge that interpretative nature of experience presents to knowledge. For standard views, the object of the sign is continually in danger of being lost, and reality along with it. Such a tenuous connection to the world typically stems from background assumptions about nature and identity that severely limit the kinds of social or relational participation involved in the careers of ordinary things. Specifically, causation is not understood as a kind of participation of one thing in the identity of another; instead, things are "simply located" or sealed off from one another, and their causal interactions are purely external. If this world of isolated entities is our background assumption, then the problems posed by interpretation are only symptoms of a much deeper incoherence in our picture of nature.

Neville argues for a different kind of a starting point: a world whose rich networks of causal participation present infinite possibilities for engagement in experience. In such a world, interpretation is an inescapable part of experience not because "things-in-themselves" are cut off and inaccessible, but because things are so multifarious in their mutual entanglements that engagement must always be selective if it is to be meaningful. Thus, following Peirce, Neville argues that the signs of experience selectively engage certain respects of things as relevant for some particular context of interpretation, which means that the relevant dyadic relation is dependent on this entire context. Accordingly, whether or not experience "says" something true of its object depends on an assertion about the relevant *respect* in which a sign is mediating its object for a particular situation. Thus you might say that, for this view, experience *always* involves *some* grasp of reality; the trick is to gain proper control of that grasp such that it can guide us appropriately.

The discussion of valuation in preceding sections has prepared us to view the selective mediation of signs as an instance of valuing. Indeed, in its broadest application, the Peircean theory of semiosis is a theory of causation qua valuation, although it is not usually presented in these terms. What Peircean semiotics brings to the current analysis is the suggestion that valuing enacts a triadic structure: valuing is an interpreting activity (1) that takes the object or thing valued (2) to be valuable in some way by enabling it to be included as a component in a new value (3). This mixture of triadic semiosis and valuation may seem forced, but it has certain advantages over these two components taken separately.

Viewing the triadic pattern of semiosis as an act of valuation helps to bring out the former's dynamic character as a selective process, which is often lost in schematic representations of Peircean semiosis that suggest three independent entities connected by timeless relations. The third element of semiosis, the interpretant, is not the interpreter but rather the interpretive "decision" that takes the sign as mediating the object in a certain respect. It is possible to abstract from an interpretant a general habit of interpretation and consider the logical relations of this habit as if they were timeless, but we should not forget that such relations are embodied in semiosis by an event, a particular *taking* of a sign in a certain way. The dynamic character of semiosis is further accentuated in instances of *intentional* interpretation: an intentional interpretant takes up the object into a context shaped by the agent's purposeful activity within a particular situation. In such cases of intentional interpretation, the mediating activity of semiosis is "directed toward" the object by the interpretant so as receive the object in a particular way, just as described in our earlier discussion of neurodynamics. The recurring theme of this argument is that this "stretching forth" so as to receive the object (Freeman 1999, p. 27) has the character of valuation.

From the other side, viewing valuation as containing a triadic relation highlights a crucial point: valuation always entails something valued as its object. Of course, the nature of valuation is to modify that thing, selecting some features for emphasis and inclusion within the broader context that constitutes the perspective of valuation. With respect to the thing valued this modification might be a drastic simplification, distortion, reduction, or trivialization. But just as the triadic structure of Peircean semiosis prevents the object from being subsumed by the sign, an act of valuation cannot be identified as such if the act of valuation completely subsumes the thing valued. On the contrary, valuation presumes that the value of the object (2) is not exhausted by its participation in the new value (3) constituted by the act of valuation (1). Keeping the triadic nature of semiosis in mind, then, helps to prevent the total collapse of valuation into subjectivism, for which value exists only for the valuing perspective.

And this brings us to what is perhaps the most important claim of Neville's argument, namely that the thing valued as the object of interpretation is a value in its own right. "Valuation, on this hypothesis, is when the value of something becomes a value or plays a value-laden role in something else" (p. 63). That is, an act of valuation constitutes a *new value* in which the value-identity of the object plays some kind of participatory role. The claim that value is always an ingredient in valuation and not just its product means that value cannot be reduced to mental acts of valuation. The world we encounter is a world of values: the values of experience are the values of the world as discovered and revealed to us through the forms of experience. How does this view square with our experience of value?

As discussed in the previous sections, there are reasons inherent to the nature of value that prevent us from being able to demonstrate its role in experience as clearly as we would like. Moreover, the prevalence of subjectivist discourse about value—in the sciences as well as the humanities—has biased us against the notion that value is something encountered in the world. In the face of such bias, a value realist may be tempted to overstate the evidence for intrinsic value. For instance, to argue for intrinsic value one might try to amass examples of eminent worthiness, such as classic works of art. Powerful as such works may be, when they are rounded off and claimed as objects of intrinsic value the evidence can always be gainsaid. Even if these objects were universally appreciated, that might only serve to indicate universal human preferences. The better move, then, is to look at the experience of value as a dynamic process and consider the most plausible conditions for our experience of value as something *discovered*. From this point of view, it is possible to argue—though, again, not as conclusively as we would like—that the experience of value as something endlessly discoverable does not make sense unless it begins with things whose valuable character is not subsumed by their value for the subject. Our experience of the discovery of value seems to indicate that value is something that experience works *with*, both raw material and product. That is, experience does not turn "mere facts" into values, like spinning straw into gold: rather it is a process of bringing together certain values into new arrangements with other values.

> The value we give something for ourselves is always some compromise between the value it presents and our need to accommodate it to the other values. Some people might argue that a thing is just a fact with no value of its own, and that we give it a place in our experience solely on the basis of our own interests. But how could it have any bearing on our interests, how could it contribute anything (or its exclusion make something else worthwhile possible), if it did not have a valuable character itself? Where would our own interests come from, our own emotions, tastes and imaginative fancies, if not from the combined values of our conditions? To deny that things are experienced as bearing value is to deny what confronts us at all levels of personal and social life. It is to deny that poetry is true (p. 135).

Still, the fact that experience never reaches beyond valuation to a valueless world does not by itself warrant the bold speculative extension of value and valuation to all natural processes. Thus Neville devotes much of *The Recovery of the Measure* to an argument for value as an intrinsic character of all things, including an argument for an axiological metaphysics as well as a philosophy of nature that lays out the role of value in causation and the flow of time. Here we will only consider how the upshot of this argument might be brought to bear on the challenges presented by neurodynamics for realism and truth.

In Neville's words, the central thesis of his argument is that "truth is the properly qualified carryover of value of a thing (or situation, or state of affairs, or fact…) into interpreting experiences of that thing" (p. 65). "Carryover" is Neville's word for the causal participation, or presence, of the object in experience. "Properly qualified" signals that the thing interpreted is present as modified, that is, as "objectified," within the experience of the interpreter. But the crucial claim is that a thing qua object can be multiply located within the interpretive process and carried over again into subsequent interpretations. Moreover, the key move in support of this claim—what makes Neville's theory a peculiar twist on classical realism—is the argument that this carryover should be understood in terms of value rather than form.

Whereas Neville elaborates his theory of value in terms of harmony, for the sake of simplicity and continuity with previous discussion we shall focus here on a single characteristic of harmony: contrast. It may seem rather devious—a "bait and switch"—to shift from value to contrast at this point of the discussion. Yet I hope that previous discussion has prepared this move by indicating the close relationship between contrast and value. We have already considered at length the possibility that perceptual categorization, qua valuation, results in a meaningful contrast: that is, a pattern of activity whose meaning depends on the potential consequences of its differentiation of a "contrast space" (qualitative state space). Modifying the terms of Neville's thesis, then, we could say that such contrasts are true if they carry over the appropriate contrast, properly qualified, from their intended object into the experience of the organism. But are contrasts also resident in things (and not just in our experience of them)?

Earlier we seemed to deny this, observing that colors and other perceptual contrasts are "brought forth" by perceptual neurodynamics. But what is the interpretive purpose of such contrasts in perception? Perceptual contrasts are false if we take them as special kinds of "feature detectors" that are supposed to correspond to specific properties like surface reflectance. But it is evident in the case of color vision and olfaction that perceptual contrasts are *not* feature detectors. Moreover, perceptual contrasts themselves are not what we perceive (except when we focus on sensation rather than perception), rather they are *what we perceive with*: they are embodied *forms of engagement* used to pick out and identify objects so that they can be handled in some way.

Now the thesis of axiological metaphysics is that the things that we engage as objects through perceptual contrasts also constitute contrasts, albeit in different forms. And so here is the crux of the matter: to play this crucial role in experience, contrasts must be embodied by the subject in some concrete form that is different from the forms of the things engaged in experience. Thus, for carryover to be possible, it seems that a contrast must be more abstract than the particular form that embodies it. So is contrast just a more abstract level of form (like the roundness of a circle)?

If contrast were just a more abstract level of form, then the thesis under consideration would only preserve the possibility of truthful correspondence at a vague level—experience could only be "sort of true." But what seems to be the case is that we use highly detailed, multidimensional contrasts to engage highly complex situations. This level of detailed engagement seem to throw us back onto the option that we have already discarded, namely a close and regular mapping between formal features of experience and formal features of the world. To overcome these two options—highly abstract correspondence, on the one hand, and detailed formal correspondence, on the other—we need to consider that although that the carryover of contrast does involve abstraction and re-specification with respect to form, it is not merely a process of formal translation.

Recall that the carryover of contrast in experience is a selective process guided and normed by particular purposes and interests: it is nested within a context of activity, and its purpose is to guide that activity toward some end in view. If the nature of truthful experience is not to mirror the world but to guide our interaction with it, the contrasts that it carries over should serve to shape our activity so as to deal appropriately with the situation at hand. What kind of contrasts should be carried over from a thing or situation into our interactive behavior with that thing? It seems that what we need is more of a pathway or guideline than a map (cf. Ingold 2000). The kind of "fit" that we seek between activity and situation (directed toward a particular end in view) is a way of orienting and organizing that activity so as to make it appropriately responsive as it co-evolves with the situation—that is, as it develops over time and within a context of continual (or at least recurrent) engagement. Engagement means that experience is somehow geared to its objects in a manner that is much more like coordinated movement than mirroring.

Accordingly, the kind of thing that needs to be carried over is not just form, or an abstract level of a form, but rather something intimately related to form: the organizing role that form plays in the changing but continuous identity of a thing or situation as a valuation of diverse components. The resident contrasts of a thing are the ways in which its forms integrate the totality of its complex enduring identity. Accordingly, carrying this organizing or integrating aspect of an object's form into experience means that our activity is organized in relation to the object qua *process*. The crucial point is that insofar as things, situations, etc. are continuously reintegrating themselves as systems, their changing forms of integration over time will constitute contrasts that can be coordinated with the *different* forms that integrate experience and behavior over time.

Thus the carryover of value closely resembles the dynamic concept of "structural coupling" as developed in autopoietic theory (Maturana 2002). Structural coupling is a continuous form of causal interaction between two systems that produces an "interlocked history of structural transformations" (Varela 1979, pp. 48–49). The key difference is that while structural coupling, at least in its original definition, is a strictly symmetrical relation between transformations of formal properties, the "carryover of value" is the result of an intentional decision on the part of an experiencing subject who actively seeks to "couple" with objects in a particular way. In lieu of a more detailed comparison, let it suffice to say that the concept of structural coupling is perhaps the best available formal description of experience as valuation, yet insofar as it remains purely formal, it remains incomplete.

Understanding the carryover of value, therefore, depends on locating a subtle difference between value and form. Forms can be considered as relations without regard for the importance of the things related, whereas values always entail differentiations of importance—i.e. contrasts. So, for instance, the contrasts that constitute an enduring thing can be represented as the qualitative behavior of a complex system and thereby viewed as continuous transformations of form over time. However, while contrasts can be approximated by mathematical models, they cannot be reduced to mathematical formulae because, qua valuations, contrasts include all the "neglected" variables that are excluded by formulae. Value necessarily entails perspective, emphasis and attenuation, vagueness and triviality; form does not. Valuations are existential phenomena; forms are abstract generalities. Valuations constitute formal properties through the imposition of limit on diversity, and to identify valuations we have no choice but to refer to these forms: yet every valuation is more than the form it embodies. The simplifications of valuation are always contextualized by a more complex background that includes what is simplified; forms are these simplifications taken by themselves, without regard for context or background.

Given the intimate connection between form and value, it is not surprising that our thinking about truth and correspondence has tended to conflate them. In philosophy, the conflation of value and form is reinforced by standard examples of philosophical reflection and analysis—relatively simple entities (e.g. ball) or features (e.g. color) for which cognitive success is simply the recognition of an object as an instance of a general category. General categories are values whose characteristic formal features are relatively indifferent to the particularities of individual instances. This indifference to particularity is a tell-tale feature of valuation, but insofar as it is not recognized as such categorical values are easily conflated with their formal definitions. The difference between value and form is much easier to locate in complex interpersonal experiences—dancing, martial arts, conversation, etc.—for which cognitive success is some kind of skillful interaction. In such cases it easier to see how the relevant concepts are specified through their application in engagement with particular objects.

For example, to care properly for a person it is necessary to integrate as much of their individual identity into our experience as possible, so that their feelings are given a prominent, organizing role in our treatment of them. Surely the faithfulness demanded by the proper care of another person constitutes a kind of truthfulness, yet it is hard to see how it depends on formal correspondence. Rather, performances of skillful personal care are like the intricately coordinated and finely articulated movements of expert dancers: they are intentionally directed ways of evolving together with another person. Skillful coordination might look like formal correspondence—and in many cases it might be possible to represent this coordination mathematically (i.e. as a form of structural coupling)—but as a process of valuation it is always more than this. On the subjective side, cognition involves an intentional decision (although not necessarily deliberate) to embody some aspect of the value-constituting perspective of another in a particular way, and this valuation constitutes a pathway that, if successful, guides our interactions with that person toward some end in view. The actual pathway that results depends on both the intentional valuation of the subject and the constitutive valuations of the intended object.

These examples indicate that the subtle but crucial difference between formal correspondence and the carryover of value emerges most clearly in the midst of a process of extended engagement whose outcome is more or less open-ended. It would seem, then, that the dynamic, temporal character of experience is indispensable to understanding the role of value, and yet this is precisely the aspect of experience that is most frequently eliminated from analysis. If we cut up experience into slices of a timeless formal structure, essential features of valuation such as importance and vagueness disappear. Even the temporal dimensions of experience can be inadvertently reduced into static forms: for instance, the difference between foreground and background, which entails the future *determinability* of the vague penumbra of experience, can be converted into the difference between the proximate and distal ends of an *already-determined* sequence. In the former case the perspectival character of experience is the unique feeling of an event that is coming into being, while in the latter case it is reduced to a viewpoint within a completed structure. Probably something like this latter notion of experience—truncated, finished, and non-temporal—is what makes contrasts, and thus values, seem reducible by analysis to purely formal properties. In actual experience, however, contrasts are not static and closed but partly undetermined, open-ended structures. Values that are singled out as such are merely the "consummatory phases" of continually ongoing processes of valuation: values are both products and starting points of acts of *valuing*. In other words, values of experience are better understood as *pathways of engagement* with, rather than snapshots of, the world.

At issue here is one of the central challenges of axiological philosophy: namely, to call attention to the omnipresence and indispensability of value in experience while, at the same time, explaining how it is that value is so elusive that it has become widely regarded as somehow less than fully real. On this account, part of what makes the character of valuation so elusive is the fact that any pathway of engagement can always (at least in principle) be carved up into pieces and examined as a series of finished forms from which essential features of valuation such as vagueness and importance have been eliminated. The irony is that experience that has been denuded of value, while seemingly more transparent to analysis, leads to the now-familiar quandaries of modern epistemology. The commonsense expectation is that the forms of experience should correspond to relevant snippets of the world; then, when this expectation is frustrated, the temptation is to conclude that the world is "made" or "constructed" by the mind.

From the wider perspective of interactive engagement that includes valuation, however, it makes as little sense to expect that experience neatly corresponds to features of the world as to conclude that experience is entirely constructed by the mind. If we understand the mind as continually "stretching forth" so as to receive the world in a particular way, a pathway of engagement is clearly a joint venture. As such, the most plausible explanation for the rich and inexhaustible contrasts of experience is the existence of an abundant supply of values—i.e. contrasts—already resident in both body *and* world (cf. Beaton 2013). To make value solely a product of the mind—and thus valuation solely an act of certain bodily processes—leaves us unable to account for the fact that some pathways of engagement are discovered to be much more rewarding than others. Our experience of value as something discovered rather than made is a crucial piece of evidence that points toward valuation as a basic feature of the world.

10.5 Conclusion

The preceding argument has repeatedly touched on metaphysical questions of value but has stopped well short of presenting an argument for axiological metaphysics (see Neville 1989). In lieu of such an argument, I would like to close by considering the implications of axiology for standards of scientific explanation. As adumbrated here, one of the most provocative implications of axiological metaphysics is the notion that vagueness is a feature of reality and not just an illusion generated by our epistemic limitations. That is, axiological metaphysics claims that some things really are vaguely determined in relation to other things and, on at least on the surface, this would seem to present a barrier to scientific explanation. Accordingly, the initial plausibility of an axiological approach hangs on showing that, far from requiring the abandonment of rational understanding, it actually promises to make our universe *more* intelligible.

We seem to have reached a point in history where certain ideals of scientific explanation—complete intelligibility and complete knowability—are increasingly at odds with another fundamental criterion, adequacy to experience. Nowhere is this more apparent than in the field of neuroscience as applied to the human being. Like many other philosophers, I believe that scientific understanding of the mind can only move beyond its current dualistic impasse if it loosens the Cartesian requirements of mechanistic reducibility and intelligibility that have served science so well over the past several centuries. The reason is that Cartesian reduction, when carried to the extreme, yields intelligibility at the expense of adequacy: as far as they go, mechanistic systems are wonderfully effective as tools of explanation, but they seem to be very limited in terms of applicability. In their eagerness to denounce Cartesian reduction, however, philosophers have not been entirely forthcoming about what it means to abandon this framework. In particular, it is important to admit that a non-mechanistic philosophy of nature will not yield for us the same thoroughgoing intelligibility that we have become accustomed to find (at least, prima facie) in standard models of scientific explanation.[5]

Moreover, philosophers need to show that this sacrifice of total intelligibility constitutes an overall *gain* for understanding, and that no arbitrary barriers to naturalistic inquiry have been erected. Philosophical anti-reductionism should not be a form of special pleading for cherished views of the human person. Rather it should be shown that a non-reductive approach allows what was previously ignored or "eliminated" to become *more* amenable to scientific inquiry, not less. In other words, it needs to be demonstrated that the abandonment of the Cartesian framework and its standard of perfect intelligibility eventually results—ironically—in a more coherent, and therefore more intelligible, world. This trade-off has been one of the main themes of the argument for experience and causality as valuation: if valuation is essential to a coherent account of experience and nature, then it is essential to the intelligibility of our place in the world; at the same time, the nature of valuation may be such that it cannot be rendered perfectly intelligible without altering its essential character. Value may be indispensable to intelligibility but *not* perfectly intelligible according to the standards established by Descartes: this is the case if value is essential to identity but cannot be reduced to clear and distinct "simples."

[5] For instance, philosophers who argue for the emergence of mental properties seem to accept Cartesian standards of explanation for "merely physical" processes. As a result, these proponents of emergence seem either to be abruptly changing the rules of the game for mental properties, or to be insisting that emergent phenomena are more intelligible than they really are.

Even so, Cartesian standards of intelligibility have become so deeply engrained within the modern scientific and philosophical mentality that even the slightest diminishment of their scope is likely to seem anti-scientific and anti-rational. The job of axiological metaphysics, then, is to help make possible the kind of philosophical sea-change that would allow us to consider the possibility that the universe is full of non-reducible entities and yet still hangs together in ways that are rational and intelligible. The key for this new perspective is to understand value as an essential component of the rational self-determination of all things.

References

Alexander, T.M. 1987. *John Dewey's theory of art, experience, and nature: The horizons of feeling*. Albany: SUNY Press.

Barrett, L. 2011. *Beyond the brain: How body and environment shape animal and human minds*. Princeton: Princeton University Press.

Barrett, N.F. 2009. The perspectivity of feeling: Process panpsychism and the explanatory gap. *Process Studies* 38(2): 189–206.

Beaton, M. 2013. Phenomenology and embodied action. *Constructivist Foundations* 8(3): 298–313.

Chirimuuta, M., and I. Gold. 2009. The embedded neuron, the enactive field? In *The Oxford handbook of philosophy and neuroscience*, ed. J. Bickle. Oxford: Oxford University Press.

Cosmelli, D., J.-P. Lachaux, and E. Thomson. 2007. The neurodynamics of consciousness. In *The Cambridge handbook of consciousness*, ed. P.D. Zelazo, M. Moscovitch, and E. Thompson, 731–772. Cambridge: Cambridge University Press.

Damasio, A.R. 2005. The neurobiological grounding of human values. In *Neurobiology of human values*, ed. J.-P. Changeux, A.R. Damasio, W. Singer, and Y. Christen. Berlin: Springer.

Dewey, J. 1958. *Experience and nature*. New York: Dover Publications.

Dewey, J. 1980. *Art as experience*. New York: Perigee Books.

Dewey, J. 1988. The philosophy of Whitehead. In *John Dewey, The later works, 1925–1933*, 14: 1939–1941, ed. J.A. Boydston. Carbondale and Edwardsville: Southern Illinois University Press.

Dewey, J. 1991. Logic: The theory of inquiry. In *John Dewey, The later works, 1925–1933*, 12: 1938, ed. J.A. Boydston. Carbondale and Edwardsville: Southern Illinois University Press

Dewey, J. 1997. *The influence of Darwin on philosophy and other essays*. Amherst: Prometheus Books.

Di Paolo, E.A., M. Rohde, and H. De Jaegher. 2010. Horizons for the enactive mind: Values, social interaction, and play. In *Enaction: Toward a new paradigm for cognitive science*, ed. J. Stewart, O. Gapenne, and E.A. Di Paolo. Cambridge, MA: MIT Press.

Edelman, G.M., and G. Tononi. 2000. *A universe of consciousness: How matter becomes imagination*. New York: Basic Books.

Freeman, W.J. 1991. The physiology of perception. *Scientific American* 264: 78–85.

Freeman, W.J. 1999. *How brains make up their minds*. London: Weidenfeld & Nicolson.

Freeman, W.J., and C.A. Skarda. 1990. Representations: Who needs them? In *Brain organization and memory cells, systems, & circuits*, ed. J.L. McGaugh, N. Weinberger, and G. Lynch. New York: Guilford Press.

Grabenhorst, F., and E.T. Rolls. 2011. Value, pleasure and choice in the ventral prefrontal cortex. *Trends in Cognitive Sciences* 15(2): 56–67.

Hubel, D.H., and T.N. Wiesel. 1959. Receptive fields of single neurones in the cat's striate cortex. *The Journal of Physiology* 148(3): 574–591.

Ingold, T. 2000. *The perception of the environment: Essays on livelihood, dwelling, and skill*. New York: Psychology Press.
Kelso, J.A.S. 1995. *Dynamic patterns: The self-organization of brain and behavior*. Cambridge, MA: MIT Press.
Maturana, H. 2002. Autopoiesis, structural coupling, and cognition: A history of these and other notions in the biology of cognition. *Cybernetics & Human Knowing* 9(3–4): 5–34.
McDowell, J. 1998. *Mind, value, and reality*. Cambridge, MA: Harvard University Press.
Neville, R.C. 1981. *Reconstruction of thinking*. Albany: SUNY Press.
Neville, R.C. 1989. *Recovery of the measure*. Albany: SUNY Press.
Neville, R.C. 1995. *Normative cultures*. Albany: SUNY Press.
Noë, A. 2009. *Out of our heads: Why you are not your brain, and other lessons from the biology of consciousness*. New York: Hill and Wang.
James, W. 1983. *The principles of psychology*. Cambridge, MA: Harvard University Press.
Jones, J. 1998. *Intensity: An essay in Whiteheadian ontology*. Nashville: Vanderbilt University Press.
Juarrero, A. 1999. *Dynamics in action: Intentional behavior as a complex system*. Cambridge, MA: MIT Press.
Rohde, M. 2010. *Enaction, embodiment, evolutionary robotics: Simulation models for a post-cognitivist science of mind*. Paris: Atlantis Press.
Rorty, R. 1979. *Philosophy and the mirror of nature*. Princeton: Princeton University Press.
Schulkin, J. 2008. *Cognitive adaptation: A pragmatist perspective*. Cambridge: Cambridge University Press.
Smolin, L. 2013. *Time reborn: From the crisis in physics to the future of the universe*. New York: Houghton Mifflin Harcourt.
Sporns, O. 2011. *Networks of the brain*. Cambridge, MA: MIT Press.
Strawson, G. 2006. Realistic monism: Why physicalism entails panpsychism. *Journal of Consciousness Studies* 13(10–11): 3–31.
Thompson, E. 2007. *Mind in life: Biology, phenomenology, and the sciences of mind*. Cambridge, MA: Belknap.
Varela, F.J. 1979. *Principles of biological autonomy*. New York: Elsevier North-Holland, Inc.
Varela, F.J., E.T. Thompson, and E. Rosch. 1991. *The embodied mind: Cognitive science and human experience*. Cambridge, MA: MIT Press.
Whitehead, A.N. 1967a. *Science and the modern world*. New York: Free Press.
Whitehead, A.N. 1967b. *Adventures of ideas*. New York: Free Press.
Whitehead, A.N. 1978. *Process and reality: An essay in cosmology*, corrected ed. D.R. Griffin and D.W. Sherburne. New York: Free Press.

Chapter 11
Ethics and Normativity

John Cottingham

11.1 Morality: Why Darwinian Deflationism Fails

Many modern ideas about the nature of morality are strongly influenced by the kind of historical or genealogical approach that began to emerge in the nineteenth century. Perhaps the most famous proponent of this was Charles Darwin. In his *Descent of Man* (published in the 1870s) Darwin displays what can be called a 'reductionist' attitude to human morality: instead of providing us with insight into ultimate meaning and value, our faculty of moral judgement is simply a product, or by-product, of how our ancestors happened to have evolved in the struggle for survival.

In the course of Chapters 4 and 5 of the *Descent of Man*, which are about the evolution of our moral sensibilities, Darwin uses a highly significant phrase – the '*so-called* moral sense'.[1] His reductionist approach sees conscience, and other so-called 'higher' impulses, as merely one or more of many natural feelings that have developed under selection pressure. Altruism and self-sacrifice, for instance (to take an example he discusses), may have arisen because tribes in which this trait is prominent 'would be victorious over most other tribes, and this would be natural

This paper was presented as the opening address at the XLVIII Reuniones Filosóficas Conference on *Biologia y Sujectividad*, University of Navarra, Spain, April 2011, and I am grateful to participants at the Conference for helpful discussion of the arguments. At various points the paper draws on material which I have developed in previous works, details of which are footnoted below.

[1] '… actions are regarded by savages … as good or bad, solely as they obviously affect the welfare of the tribe … The conclusion agrees well with the belief that the so-called moral sense is aboriginally derived from the social instincts…' Darwin [1871; 2nd. 1879] (2004), Ch. 4, p. 143.

J. Cottingham (✉)
University of Reading, Reading, UK
e-mail: jgcottingham@mac.com

M. García-Valdecasas et al. (eds.), *Biology and Subjectivity*,
Historical-Analytical Studies on Nature, Mind and Action 2,
DOI 10.1007/978-3-319-30502-8_11

selection'.[2] The crucial point Darwin is making here is about the purely natural origin of our moral feelings; they are in this respect just like any other ingrained drives and inclinations – part of our natural, animal inheritance. Indeed, he implies that the difference between moral and non-moral feelings is rather like the difference between human dispositions and those occurring in other primates: the difference may be considerable, but in Darwin's view it is 'certainly … one of *degree,* and not of kind.'[3] There is nothing special, nothing specially exalted, about morality or about the conscience or moral sense that supposedly detects moral values. The entire phenomenon of morality can take its place, in principle, as simply another part of the natural world.

This classic Darwinian position is a bold one, but it is faces serious problems. For when we normally think of the domain of morality it does appear to differ in kind from anything found in the ordinary observable natural world.

Moral values, to begin with, appear to be *objective* – they do not seem to be a function of my personal preferences and desires, or even those of society in general. Cruelty and arrogance are objectively wrong, and remain so irrespective of whether I have a taste for them. Even if arrogance became universally admired, that would not show it was right or good,[4] only that human beings had become more corrupt (something that is, of course, all too possible).

Second, fundamental moral values are *universal* – not in the sense that they are as a matter of fact respected always and everywhere, but in the sense that their validity and scope is independent of the variations of local history or geography. Conceptions of virtue do of course differ in different epochs and tribes – something that Darwin highlights; but that does not count decisively against objectivity and universality. The wrongness of slavery, for example, or the goodness of compassion, may not be universally acknowledged in all ages and areas, but that does not prevent their reflecting perfectly objective and universal truths about virtue and value. After all, the truths of physics are very far from universally acknowledged, but that does not at all count against their being universal and objective (it is just that they take a lot of time and effort to uncover and understand properly).

Thirdly, moral values are *necessary* – cruelty does not just *happen* to be wrong, but is wrong in all possible worlds. We may of course transgress such fundamental norms, and often do, but (as the nineteenth-century logician Gottlob Frege put it in

[2] 'A tribe including many members who … were always ready to aid one another, and to sacrifice themselves for the common good, would be victorious over most other tribes; and this would be natural selection. At all times through the world tribes have supplanted other tribes; and as morality is one important element in their success, the standard of morality and the number of well-endowed men will thus everywhere tend to rise and increase.' Darwin [1871; 2nd. 1879] (2004) Ch. 5, pp. 157–58. Modern evolutionary theorists would see this apparent endorsement of group selection as problematic, but, with the aid of genetic theory, could easily adjust the story, rewriting in terms of the advantages of prevalence within a given population of an individual gene or genes linked to altruistic behaviour.

[3] Darwin [1871; 2nd. 1879] (2004), Ch. 4, p. 151.

[4] There are of course important differences between 'right' and 'good', but I shall not be exploring them in this paper, since they do not affect the main thrust of my argument.

a rather different connection, discussing the truths of logic and mathematics), they are rather like 'boundary stones which our thought can overflow but not dislodge' (Frege [1893] (1964), p. 13).

And fourthly and finally, moral values are, in the current philosophical jargon, *normative* – that is to say, they exert an authoritative demand or call upon us, whether we like it or not. This last is a remarkable property, which what used to be called the 'eternal values' (truth, beauty and goodness) all share – they carry with them the sense of a *requirement* or a *demand*. Some languages, Latin for example, have a special grammatical form, called the gerundive, to express this notion. Thus, the Latin word *amandus* (from *amare*, to love) means not just that something is loved, or even that it is 'lovable' (in the rather weak sense that it tends to be loved, or is apt to be loved), but rather that it is *to be loved*, that it ought to be loved. This kind of 'gerundive' flavour seems to attach to truth: the true is that which is worthy of belief – *to be believed*. And similarly the beautiful is that which is worthy of admiration, *to be admired*, and the good is that which is worthy of choice, *to be pursued*. Truth, beauty and goodness therefore seem to be rather 'queer' properties (as the philosopher John Mackie once put it: 1977, Ch 1, §9): they have this odd, magnetic aspect – they somehow have 'to-be-pursuedness' built into them. Why is this queer or odd? Because it seems incompatible with any purely *naturalistic* or scientific account of these properties; for it is not easy to see how a purely natural or empirically definable item could have this strange 'normativity' or choice-worthiness somehow packed into it. So it starts to look as if thinking about these normative concepts is sooner or later going to take us beyond the purely natural or empirical domain.[5]

Some philosophers think that this is making a lot of fuss about very little. For example, we have recently seen the rise of so-called 'buck-passing' accounts of value. The American idiom, to 'pass the buck', means to shift responsibility to someone else. So 'buck passing' accounts of value shift the emphasis away from normative notions like 'good' and 'ought', onto much less puzzling and more down-to-earth factual notions that seem to underpin them. What makes a knife good is simply that it has certain ordinary natural or observable features (such as sharpness) which give me reason to choose it, if I want to cut something. And similarly, what makes charity good, for example, is simply that the relevant actions have certain ordinary natural properties (e.g. reducing suffering) that give me reason to perform them (for the 'buck-passing' account, see Scanlon 1998).

Yet although it is obviously true that goodness and badness are connected with ordinary features of actions in this way (the ordinary observable features providing *reasons* for us to choose or avoid things), it is unfortunately all too clear that many people are not responsive to such reasons. Many people delight in cruel or vicious behaviour; and the suffering of others that *we* may regard as a reason for them to desist, simply is not recognized by them as a reason to stop. We may say: 'Yes, but whatever such vicious people may say or feel, the suffering of their victims *is* a

[5] For more on this and the other aspects of moral value discussed here, see Cottingham 2009, Ch. 2 and 2005, Ch. 3.

reason, an *authoritative* reason, for them to desist'. But then we seem to be appealing to some kind of moral demand that remains in force *no matter what* – no matter how many people transgress it, or refuse to recognize it. In short, although the 'buck-passing' account of value seems right in grasping how goodness and badness point beyond themselves to ordinary natural features of actions that provide reasons to choose or avoid them, it does not appear to explain how some of those reasons possess *authoritative normative power*.

The conclusion of the argument so far is that the Darwinian or naturalistic approach to ethics is in trouble when it comes to explaining genuine moral values—genuine in the strong sense I have been discussing, which is manifested in the fact that they provide conclusive reasons for action whether we like it or not.

11.2 Conscience and Its Origins

Now there is an obvious way of trying to block this conclusion: one could just deny that there are any values of this sort: one could say that the traditional notion of eternal values is a fantasy, or that the whole idea of an objective, universal, necessary, normative standard is an illusion, or a sham. Darwin himself is too cautious to say this outright, but he does frequently note the findings of anthropologists and explorers about variations in moral norms from culture to culture, and indeed the total absence of some of our most cherished values in what the Victorians called 'savage' peoples.

The general message from *The Descent of Man* is that there is no ultimate court of appeal in the face of such variations. Instead of a moral absolute called 'goodness', we simply have an efficiency criterion; when we talk of the 'standard of morality', the only serious scientific test can be 'the rearing of the greatest number of individuals in full vigour and health, with all their faculties perfect, under the conditions to which they are subjected.' (Darwin [1871; 2nd. 1879] (2004) Ch. 4, p. 145). On this *deflationary* view,[6] we human beings just invest our moral values with a kind of mystique, an aura of power; but they are simply human inventions, or projections from the desires and drives that we happen to have evolved in the struggle for survival. Of course, there are social rules and norms – that is a matter of history or social anthropology. But these (so this view holds) are reducible to natural phenomena – learned behaviour patterns, instilled social practices – which the empirical scientist can investigate and describe. This is the deflationary line that Darwin took when he meditated on 'the imperious word *ought*'. 'The imperious word *ought*', he says in the *Descent*, seems merely to imply the consciousness of the

[6] Imagine a magnificent balloon floating in the air. If it is *deflated* it falls to the earth, no longer an object of wonder and admiration. So by calling an account of morality 'deflationary', I mean that its effect is to take away the sense of moral values as having a exalted power or authority, and present them instead as something much more ordinary and down to earth.

existence of a rule of conduct, *however it may have originated.*' (Darwin [1871; 2nd. 1879] (2004), Ch. 4, p. 140).

A decade or so earlier, John Stuart Mill, writing from an entirely secular perspective, in his essay *Utilitarianism*, had taken a similar purely empirical or naturalist line. He defined 'the essence of conscience' as 'a feeling in our own mind; a pain more or less intense, attendant on violation of duty'. He adds various qualifications – that the feeling must be 'disinterested', and connected with the 'pure idea of duty' – but the main effect of his account is a deflationary or demystifying one: to reduce the deliverances of conscience to nothing more than a set of psychological events or purely subjective feelings. These feelings, he observed, are typically 'encrusted over with collateral associations', derived from the 'recollections of childhood' and 'all the forms of religious feeling'; and this, he claims, is enough to explain away 'the sort of mystical character which … is apt to be attributed to the idea of moral obligation.' (Mill [1861], Ch. 3).

Mill's account purports to be simply a piece of empirical psychology, but it clearly has serious implications for the normativity of conscience. Painful feelings, linked to the violation of duty, function as what Mill called 'internal sanctions'; and he wished to enlist these in the service of his own utilitarian ethics. But sanctions, as understood by Mill, are just causal motivators – means whereby a desired code may be inculcated into the population so as to reinforce allegiance; we are thus in the territory of *inducements for compliance*, not in the territory of *authoritative reasons for action*. Mill was sensitive to the objection that if what restrains me from wrongdoing is 'only a feeling in my own mind', one may be tempted to think that 'when the feeling ceases, the obligation ceases.' But he confines his reply to observing that those who believe in a more exalted and objective source of obligation are just as likely to transgress morally as those who think that what restrains them 'is always in the mind itself' (Mill [1861], Ch. 3). This may be true, but it is hardly relevant to the question at hand: does conscience have genuine authoritative power or not?

For the naturalist (by which I mean someone who holds that there is no reality beyond the set of natural events and processes that have emerged from the Big Bang), it is hard to see how the answer can be anything other than 'no'. In a neutral, Godless universe, where the planet and all its inhabitants are, in the end, simply accidental and temporary by-products of the debris flung out by the Big Bang, and where human impulses of conscience are simply a subset of many conflicting impulses that we happen to have evolved in the struggle for survival, it is hard to see how any one human faculty can have the special status of being (to borrow the phrase of Joseph Butler from the eighteenth century) 'a faculty *in kind and in nature supreme over all others*, and which *bears its own authority* of being so.' To be sure, we do as a matter of fact have natural dispositions to kindness and compassion. But as Joseph Butler observed, it is equally true that '… other passions [such as anger] … which lead us … astray, are themselves in a degree equally natural, and often most prevalent …[and hence] it is plain the former considered merely as natural … can no more be a law to us than the latter.' (see further Cottingham 2004, pp. 11–31). We are still far short of the traditional idea of conscience, which Butler aptly characterises as the 'superior principle of reflection or conscience in every man … which pronounces some actions to be in themselves just, right, good; others to be in themselves evil, wrong, unjust.'[7]

The strong kind of authority or normativity that Butler sought was (in his view) to be found within a theistic metaphysics. Since his time, of course, especially in recent years, there has been much philosophical energy devoted to trying to show there can be normativity in a purely natural universe. But in my view all that this recent work has delivered, in the moral or practical arena, is *conditional* or *relative* normativity: the idea that certain natural features of things provide reasons to choose or commend them *given that* we have certain projects or purposes, or *relative to* certain contingently evolved biological or social structures. If the purposes, or the structures, were different, then the normativity would evaporate, or point us in another direction.

It is this background of *contingency* that seems to me to undermine the traditional idea of strong normativity, mirrored in the authoritative intuitions of conscience. In the secular worldview, in which a random or accidental chain of events gives rise to a certain type of featherless biped with certain contingently evolved desires and inclinations and communal practices, ethics will be subject to what the distinguished moral philosopher Bernard Williams called a '*radical contingency*', by which he meant that 'our current ethical conceptions … might have been different from what they are', and that the conditions which brought them about are 'not related to them in a way that vindicates them against possible rivals'.[8] Had the evolutionary history of the planet, or our species, been slightly different, then our morality might well have been slightly or even radically different.

But notice the disturbing implications of this idea. If our ethical conceptions are a product of a purely contingent concatenation of events, if they might have been otherwise, then it starts to look as if they might be changed or overridden. As Friedrich Nietzsche put it, in the *Genealogy of Morals* (published not too long after Darwin's *Descent*), once we start to think about the conditions under which humans invented the value judgements good and evil, we can start to ask *what value do these value judgements themselves possess.*[9] It is no accident that Bernard Williams's conception of ethics, and his scepticism about what he called 'the morality system', were strongly influenced by Nietzsche, and his idea that, once we accept that ethics has a genealogy, a contingent history, this frees us from acknowledging the authority

[7] Butler [1726] (1990) and also in Raphael 1969. For more on the philosophical issues arising from the notion of conscience, see Cottingham 2013.

[8] '[A] truthful historical account is likely to reveal a radical contingency in our current ethical conceptions. Not only might they have been different from what they are, but also the historical changes that brought them about are not obviously related to them in a way that vindicates them against possible rivals.' Williams 2002, Ch. 2, p. 20.

[9] 'Fortunately I have learnt to separate theology from morality and ceased looking for the origin of [good and] evil *behind* the world. Some schooling in history and philology, together with an innate sense of discrimination with respect to questions of psychology, quickly transformed my problem into another one: under what conditions did man invent the value-judgements good and evil? *And what value do they themselves possess?*' Nietzsche [1887], Preface, §3.

of so called eternal moral values. As Nietzsche put it, we can, if we are strong enough, decide to *invert* eternal moral values. In a godless universe, where God is 'dead', then we are not subject to any higher authority, and so questions of value become merely a function of the projects we autonomously decide to pursue. So (as Nietzsche frighteningly suggests in one of the most disturbed and disturbing passages in Western philosophy), why should we not cut ourselves off from 'herd-animal morality' with its 'sympathy for whatever feels, lives suffers … in [its] almost feminine incapacity to remain spectators of suffering, to *let suffer* …' For there might (in Nietzsche's way of thinking) be conclusive reasons to steel ourselves *against* impulses of love and mercy, to harden our hearts *against* compassion and forgiveness, since such sentiments might get in the way of our will to power, or our passion for self-realisation, or some other grand project we happen to have.[10]

11.3 Is Contingency So Damaging?

The conclusion towards which I am moving, then, is that (broadly) Darwinian or naturalistic accounts of ethics are unstable because of what they imply about the contingency of ethics.[11] The contingent origins of our moral intuitions, their emergence out of a developmental flux that does nothing to vindicate them, and the realization that they might have been, and might still be, challenged and changed – this cluster of connected worries seems to undermine their authority. But let me close by considering an objection to this conclusion: is contingency really so damaging?

In answering this question, I should first make some important concessions to the naturalistic framework for understanding ourselves, which, in my support for the idea of 'eternal values', I might seem to have been rejecting altogether. First, we humans are, to be sure, creatures who belong within the natural world, and any plausible account of human nature needs to acknowledge this. Darwin was surely right to insist on this point. Human beings are indeed formed 'out of the dust of the earth' (as the second chapter of Genesis puts it), and we have to understand ourselves as part of the vast natural process of the cosmos. In the second place, it is important to note that contingency need not imply a chaotic flux. Notwithstanding the pressure that Darwinian ideas put on the idea of a fixed and immutable human nature, any plausible developmental account of our origins will allow that there are stable features of the human condition that remain largely unchanged across vast periods of time. These stabilities are of course reflected in the ethical domain. For example, Aristotelian ethics aimed at specifying those excellences of character that enable us to flourish as the kind of creatures we are – possessed of drives and needs we share with other animals, yet also having the capacity for rational reflection; and

[10] See Nietzsche [1886], §37.

[11] The final two sections of this paper draw on material from my articles: 2012, pp. 233–254; 2008, Ch. 2, pp. 25–43.

it is striking how much of that ethics continues to speak to us today. Of course it is not entirely immutable: there may be room for dispute about which virtues need to be added to or subtracted from the list; but despite all ways in which our lives diverge from those of Classical Greece, there is ample evidence from literature and history and biology to believe our human nature has changed very little, if at all, since those days; indeed, in evolutionary and genetic terms, the whole human story since prehistoric times is the merest blink of an eye.

So despite the contingencies and vicissitudes of our evolutionary and cultural history, it is perfectly plausible to maintain that any account of human flourishing must be anchored in certain relatively stable, basic facts about human nature, and that, whatever the variations in the human situation from epoch to epoch, or culture to culture, there will necessarily be a vast amount in common. So, to come to the point at issue, *why not just live with the contingency?* Why shouldn't our ethical and intuitions simply reflect certain admittedly contingent but nonetheless relatively fundamental and relatively stable facts about our biological and social nature as it has evolved over time? Nietzsche's claim, as we have seen, is that becoming aware of morality's contingent origins destabilizes it. But why shouldn't an objector accept the contingency and simply point out the *advantages* of the morality system that has emerged from the evolutionary and historical flux. For example, why shouldn't the traditional (contingently evolved) morality system gain its authority from being the set of values that, as a matter of fact happens to best serve the interests of humanity?

To answer this, I cannot do better than quote from Christopher Janaway in his excellent recent study of Nietzsche's *Genealogy of Morals*.

> Those conceptions of humanity's best interests that show morality in a strong light tend to be those that Nietzsche has argued to be part of the very same historical construct. For example, morality might be said to benefit the greatest number of people by its potential to protect them from some degree of suffering … But, when we have read the *Genealogy*, we may be persuaded that many of the constituent assumptions here – that suffering is something in principle lamentable about life, that well-being consists chiefly in the absence of suffering, that the well-being of all humans matters equally, that values are preferable the greater the number they benefit – are all part of the same elaborate, contingent body of ideas and attitudes that is morality. (2007, p. 248).

In other words, once the contingent genealogy of morals is accepted, then the authority we accord to our traditional moral intuitions will turn out to be contingent on our allegiance to the values of the morality system which our culture has developed— and yet that is the very allegiance that Nietzsche's arguments have called into question.

11.4 The Remaining Options

How might contemporary philosophers respond to these problems? Interestingly enough, much contemporary Western ethics is nowadays firmly objectivist, and committed to normative realism: it wants to retain the traditional idea of objective normative values, that are not simply a function of contingently evolved preferences or projects. But does it have the intellectual or ontological *resources* to support this strong normative realism?

I have already referred to the 'buck-passing' account, which holds that goodness is the second-order property some action or object has of providing reasons for choosing it in virtue of its natural, first-order properties. But I think this route will only provide *conditional* normativity. It will not yield the kind of authoritative demand that the intuitions of conscience were traditionally supposed to track— a call that appears to draw us forward and beyond ourselves, beyond the flux of our contingent inclinations and projects, beyond the bundle of traits and characteristics we happen to have evolved to have, towards something absolute and unchanging that demands our allegiance.

A somewhat different view that has many modern supporters is what might be called *strong realist non-naturalism*. This insists that values have a genuinely objective status: they are not reducible to natural facts and properties (hence the term 'non-natural'), but exist as normative realities in their own right. Thus Russ Shafer-Landau tells us that values are 'a brute fact about the way the world works'; or, in a later formulation, 'moral principles are as much a part of reality as … the basic principles of physics'.[12] Yet on this 'brute fact' view, are we supposed to think (to put it very crudely) that values somehow float around, alongside planets and stars and galaxies? It is one thing to say values exist, but *how* do they exist?[13] In fairness, Shafer-Landau goes on to concede that his theory is one with 'very limited explanatory resources' (Shafer-Landau 2003, p. 48). But in that case, the danger is that it will not come down to much more than the mere assertion that moral values really (mysteriously) exist.

Another strong realist, Eric Wielenberg (*Value and Virtue in a Godless Universe*) asserts that moral truths are 'part of the furniture of the universe', and indeed constitute the 'ethical background of every possible universe.' (2005, p. 52). This latter phrase suggests that we should think of values as purely abstract objects, perhaps rather like triangles or prime numbers. So if we are prepared to accept that abstract mathematical entities exist (waiting to be discovered and investigated by mathematicians), could we not perhaps accept that abstract values exist (waiting to be investigated by moralists)? Yet this kind of approach seems to invoke one mystery (the existence in all possible worlds of objective mathematical realities) in order to explain another (the existence of moral realities). If eternal mathematical and logical and moral reality is somehow involved in the very existence of things, yet cannot be explained in naturalistic terms, then this is a remarkable fact (and, one might add, remarkably consistent with traditional theism); it seems the non-naturalist needs to *respond* to this, instead of just asserting that such realities are 'part of the universe'.

---

[12] See Russ Shafer-Landau 2003, pp. 46, 48: moral standards 'just are correct'; they are 'a brute fact about the way the world works'. For the comparison with physics, see 'Ethics as Philosophy: A Defense of Ethical Non-naturalism', in Shafer Landau (ed.) *Ethical Theory*, Ch. 8.

[13] The so-called 'redundancy' strategy, of construing truth and reality claims as merely emphatic asseverations of the propositions they refer to – so that '*x is true/really true/part of reality*' is merely a strong way of asserting *x* – is of course not available to the non-naturalist, on pain of retreating from the very normative realism that he is supposed to be propounding.

As we noted earlier, the traditional approach to ethical objectivity (for example in Samuel Butler) was a religious one. And if we wish to retain the idea of a faculty which gives us insight into an objective, universal, necessary domain of value that has normative authority over us, then the traditional theistic framework at least offers a world-picture in which normativity is, so to speak, *comfortably at home*. The theist's universe, after all, is, from first to last, a *creation*, already replete with meaning and value and intelligence. This may not solve any explanatory puzzles in the way a scientific explanation does, but then neither does a theistic view of anything else. As the distinguished theologian Herbert McCabe observed, 'to say that God created the world is in no way to eliminate the intellectual vertigo we feel when we try to think of the beginning of things.' (2009). And so it will be for value. For there is a certain vertigo in the idea that we contingent creatures— 'imbecile worms of the earth', in the phrase of Blaise Pascal— are called to transcend ourselves. As Pascal observed: 'humanity transcends itself'. It is our nature always to strive to close the moral gap between what we are and what we are called to be. (Hare 1996; Pascal [1670] (1962), no. 131).

Sooner or later, as we struggle through life, we seem compelled to acknowledge, the call to orient ourselves towards values that we did not create, and whose normativity cannot be explained merely as a function of a given subset of our various naturally evolved impulses and socially evolved standards. As Shafer-Landau puts it (and here I would wholeheartedly agree with him), 'we humans have created for ourselves a number of different sets of conventional moral standards, but these are never the final word in the moral arena. The flaws and attractions of any conventional morality are rightly measured against a moral system that human beings did not create.' (Shafer-Landau, 'Ethics as Philosophy', p. 62).

Love, compassion, mercy, truth, justice, courage, endurance, fidelity— all these belong to a core of key values and virtues that all the world's great religions (and the modern secular cultures that are their offspring) recognize, and which command our allegiance whether we like it or not. We may try to go against them, to live our lives without reference to them, but if we are honest we cannot deny their authority over us. So if we reject the theistic picture that provides a home for such authoritative values, and if I am right that there is no satisfactory half-way house of non-naturalist objectivism, then perhaps the only viable option remaining to us would be to maintain that the commanding authority of moral values is a massive illusion. That of course could be the case. But when we focus, clearly and sincerely, on the way such values are disclosed to us in our deepest moral intuitions, such a deflationary line turns out to be very difficult indeed to sustain.

References

Butler, J. [1726] 1969. *Fifteen Sermons*. In *British Moralists 1650*–1800, ed. D.D. Raphael. Oxford: Clarendon.

Butler, J. [1726]. 1990. *Fifteen Sermons*. In *Moral philosophy from Montaigne to Kant,* II, 8, ed. J.B. Schneewind. Cambridge: Cambridge University Press.

Cottingham, J. 2004. Our natural guide…: conscience, nature, and moral experience. In *Human values*, ed. D.S. Oderberg and T. Chappell. London: Palgrave.
Cottingham, J. 2005. *The spiritual dimension*. Cambridge: Cambridge University Press.
Cottingham, J. 2008. The Good Life and the 'Radical Contingency of the Ethical'. In *Reading Bernard Williams*, ed. D. Callcut. London: Routledge.
Cottingham, J. 2009. *Why believe?* London: Continuum.
Cottingham, J. 2012. Human nature and the transcendent. In *Human nature,* ed. C. Sandis and M.J. Cain. Royal Institute of Philosophy supplement 70. Cambridge: Cambridge University Press.
Cottingham, J. 2013. Conscience. In *The Oxford dictionary of the history of ethics*, ed. R. Crisp. Oxford: Oxford University Press.
Darwin, Ch. [1871; 2nd edn repr. 1879] 2004. *The descent of man and selection in relation to sex.* London: Penguin.
Frege, G. [1893] 1964. *The basic laws of arithmetic* [*Die Grundgesetze der Arithmetik,* I]. Trans. M. Furth. Berkeley: University of California Press.
Hare, J. 1996. *The moral gap: Kantian ethics, human limits and god's assistance*. Oxford: Clarendon.
Janaway, Ch. 2007. *Beyond selflessness: Reading Nietzsche's genealogy*. Oxford: Oxford University Press.
Mackie, J. 1977. *Ethics: Inventing right and wrong*. Harmondsworth: Penguin.
McCabe, H. 2009. *God and evil in the theology of Thomas Aquinas*. London: Continuum.
Mill, J.S.[1861] 1998. *Utilitarianism*. Oxford: Oxford University Press.
Nietzsche, F. [1887] 1996. *On the genealogy of morals*. Oxford: Oxford University Press.
Nietzsche, F. [1886] 2002. *Beyond good and evil*. Cambridge: Cambridge University Press.
Pascal, B. [1670] 1962. *Pensées*, ed. L. Lafuma. Paris: Seuil.
Scanlon, T. 1998. *What we owe to each other*. Cambridge, MA: Harvard University Press.
Shafer-Landau, R. 2003a. *Moral realism*. Oxford: Clarendon.
Shafer-Landau, R. 2003. Ethics as philosophy: A defense of ethical non-naturalism. In *Ethical theory,* ed. R. Shafer Landau. Chichester: Wiley-Blackwell.
Wielenberg, E.J. 2005. *Value and virtue in a godless universe*. Cambridge: Cambridge University Press.
Williams, B. 2002. *Truth and truthfulness*. Princeton: Princeton University Press.

Index

M. García-Valdecasas et al. (eds.), *Biology and Subjectivity*, Historical-Analytical Studies on Nature, Mind and Action 2,
DOI 10.1007/978-3-319-30502-8

P

Q

R

S